Occupational Science in the Service of Gaia

*An Essay Describing a Possible Contribution
of Occupational Scientists to the
Solution of Prevailing Global Problems*

By

Moses N. Ikiugu, Ph.D., OTR/L

*Associate Professor and Director of Research
Department of Occupational Therapy
The University of South Dakota*

PUBLISH
AMERICA

PublishAmerica
Baltimore

First printing

PublishAmerica has allowed this work to remain exactly as the author intended, verbatim, without editorial input.

ISBN: 1-4241-9373-7
PUBLISHED BY PUBLISHAMERICA, LLLP
www.publishamerica.com
Baltimore

Printed in the United States of America

Dedication

This work is dedicated to my two children, Ivan and Nora, whose love motivated me to think about how contemporary human occupations may affect not only their future but also the well-being of my possible future grandchildren

Acknowledgments

The ideas discussed in this book originated from my intuitive sense that knowledge about occupational performance can be applied to solve wider societal problems than addressed by occupation therapy which is typically practiced within medical establishments or school systems. This notion occurred to me about four years ago while I was teaching at the University of Scranton in Pennsylvania, USA. Since then, many people have helped me to develop these ideas. In this regard, I would like to express my deepest gratitude to the following individuals:

Dr. Elizabeth Townsend and Colleagues at the Canadian Society of Occupational Scientists (CSOS) for inviting me to present in their 5th Annual Conference. This invitation encouraged me to explore ways in which scientific knowledge of daily occupations may be applied in organizational development.

Dr. Ruth Zemke whom I met for the first time at the CSOS conference in Vancouver for inviting and encouraging me to join the Society for the study of Occupation:USA (SSO:USA), where I could find support of like-minded scientists to help me develop my ideas. Her warmth and desire to help others grow is highly appreciated.

My colleagues, Prof. Lynne Anderson and Dr. William Anderson of the University of South Dakota (USD) for ensuring successful completion of the research study reported in Chapter 5. I would like especially to thank them for allowing me to publish the questionnaire in the appendix. This instrument was developed jointly by the three of us and I appreciate their willingness to give me permission to publish it in this book.

The department of occupational therapy at the USD, and especially the department Chair, Dr. Barb Brockevelt for making available resources necessary to conduct the survey completed by Prof. Anderson, Dr. Anderson, and me. The department financially supported the study in full. We could not have been able to complete the project successfully without this financial support.

Connie Twedt, program assistant at the USD occupational therapy department, for her assistance in ensuring accurate mailings of the questionnaires during the study.

PublishAmerica Production team for helping me through the process of publication of the book.

My friends and adopted United States parents Susanna Davila and Dick Curtis for their unwavering love and encouragement. I would especially like to thank them for thoroughly reading the initial draft of the book manuscript and giving me insightful and very detailed feedback. The feedback was very useful in helping develop a much better revised manuscript. It was due to their insightful comments, for example, that I added the study reported in chapter 7, which I think strengthened the chapter greatly.

Karen Brady, Assistant Professor of Occupational Therapy at the University of Scranton for reading parts of the book manuscript and providing useful feedback.

My friend Daniel Muhoro Karingithi for his assistance in acquiring the pictures published in chapter one (Figures 1-1 and 1-2). His generous contribution of time, photographic expertise, and effort in this endeavor is greatly appreciated.

Finally and not in any way the least, Marie Anne Ben for her love, support, and encouragement throughout the process of writing this book. I can never thank her enough for her invaluable love and support not only during the writing of this book but also in the process of completing many other projects.

Preface

In 1961, Mary Reilly, an eminent occupational therapist asserted that it is probable that the greatest idea in 20th century medicine was that man, by the use of his own hands, as directed by the mind and energized by the will, can determine the state of his own health (Reilly, 1962). Many people may not have realized at the time the full implications of that statement. Since then, it has been established that indeed, occupational performance is not only essential for human well-beings but it is part and parcel of their evolution. In stages of human history, humans were engaged in the productive occupations of hunting and gathering, and the self maintenance occupations of feeding and grooming among others. They developed tools to make their occupational performance more efficient and effective. Then, they settled down in agricultural communities and continued developing tools and eventually specializing in specific occupations such as tool making, animal husbandry, cultivation of crops, etc. From agriculture, industrialization evolved, and subsequently, other technological developments occurred throughout history until the current stage of a second industrial revolution characterized by advent of information technology emerged.

In all these stages of human evolution, occupation, ranging from hunting and gathering, animal husbandry, tool making, etc. was a means by which humans interacted with the environment, shaped it, and made it supportive of their survival as a species. In other words, occupation was and still continues to be the means by which humans integrate mind and body in action in an endeavor to adapt to their environment. This intricate relationship between human occupation and adaptation is what gives it such power in human life. It is not a wonder then, that Mary Reilly recognized the importance of accessing this power and using it for the purpose of human healing and health enhancement. Indeed, occupation has been used extensively by occupational therapists for health purposes.

However, I would like to argue that not even a fraction of the power of occupation has been tapped for the purpose of solving some of the serious problems facing humankind and the earth today. While engagement in occupations is the means by which humans have shaped the environment and made it favorable to their survival, unfortunately it is also the means by which they have seriously harmed the earth, our home, and the home of all living things in the solar system. As we have continued to shape the environment and make it home, we have destroyed vegetation cover, making it more and more difficult for the earth to sustain life. As a result, many species of life-forms have become extinct. We have also polluted the environment.

Lovelock and Margulis (1974) and Lovelock (1989, 2006) argued that the earth as a system consisting of the upper layers of rock, the atmosphere, oceans and the biosphere functioned as if it were a live super-organism. They called the system Gaia, naming it after the Greek earth goddess. By stripping Gaia of the vegetation cover, we are destroying her protective skin and wounding her. By polluting the atmosphere, we are affecting her respiratory system. Consequently, she has become ill and is now feverish (expressed as global warming). The result of this injury to Gaia has been an increase in frequency and intensity of catastrophic natural events such as storms, floods, droughts, etc. Resources needed to sustain life are becoming increasingly scarce. Consequently, humanity and all life on earth are faced with threatening problems. These include increasing poverty, extreme disparity in wealth among people, diseases whose effect is increasingly fatal among the poor, and corruption and dysfunction of human institutions that are supposed to help humanity solve its problems of adaptation. All this is happening while human population continues to grow in dramatic proportions, especially among poor people who are also the most vulnerable to the problems listed above.

If we do not do something to correct the situation, Gaia as we know her today, and all the life she supports may very soon die. Of late, issues of environmental destruction and global warming/climate change have been prominently at the forefront in the news and scientific media. Many solutions have been suggested. Hansen (2007) recommended a number of measures to curb global warming including increasing efficiency of energy usage, developing alternative energy sources that do not produce green house gasses such as Carbon Dioxide, mandating that industries install technology for capturing and sequestering Carbon Dioxide produced in their plants, providing economic incentives for companies and individuals to take environmental protection measures by taxing high greenhouse gas emissions, legislating energy efficiency standards, and disseminating accurate information about environmental destruction and global warming to the public. Similar recommendations have been made by the United Nations scientists (Hanley, 2007).

It is proposed in this book that the Gaia problems of environmental destruction and global warming/climate change cannot be solved in isolation. To address them effectively, other related global problems have to be addressed. As an example, the issue of poverty has to be confronted effectively in order to empower individuals to protect the environment. There is evidence that poor people in Sub-Saharan Africa depend exclusively on firewood for heating, which encourages deforestation, which contributes to environmental destruction (Kirkland, Hunter, & Twine, 2005). For these individuals to afford alternative sources of energy and contribute to protection of the environment, their poverty has to be alleviated. Similarly, overpopulation places a strain on available resources increasing the tendency towards environmental destruction and global warming. Therefore, global warming cannot be addressed successfully without dealing with those other issues.

Furthermore, it is proposed that since human occupational pursuits have caused our current predicament, we can use its power to shape the environment and social institutions in an effort to reverse the situation. That is the argument being propounded in this book. In part I of the book, consisting of chapters 1 to 3, some of the symptoms of the ailing Gaia are explored. These include poverty, diseases, corruption and institutional dysfunction/failure, war, overpopulation, and global warming/climate change.

In part II, chapters 4 and 5, the proposition that occupational science may be a source of a conceptual framework for use in solving the global issues that afflict humanity and other life-forms on earth is explored. The relationship between human occupational performance in the areas of work, self-maintenance, leisure pursuits, war, and sexual activity, and the global issues of concern is illustrated. In part III, chapters 6 and 7, an individual-centered, occupation-based framework for solution of the global issues is presented. In the conclusion, some ideas regarding how the conceptual framework can be applied and how its effects can be investigated through empirical research are presented. It is hoped that the ideas presented in this book will spark a spirited discussion among occupational scientists, occupational therapists, social scientists, and other professional and scientific disciplines interested in addressing global issues facing us, regarding how the power of occupational performance can be harnessed for use in solving the social and ecological issues of our time.

References

Hanley, C. H. (2007). Scientists to U.N.: "Tens of billions" needed to combat global warming. *USA Today*. Retrieved February 27, 2007, from http://www.usatoday.com/weather/climate/globalwarming/2007-02-27-scientists-un_x.htm?csp=24.

Hansen, J. (2007). *Global climate change*. Talk aired on C-SPAN2 Television on February 27, 2007. First aired on July 22nd, 2005.

Kirkland, T., Hunter, L. M., & Twine, W. (2005). *"The bush is no more": Insights on institutional change and natural resource availability in rural South Africa*. Institute of Behavioral Science, Research Program on Environment and Behavior, University of Colorado at Boulder, Boulder, CO, Working Paper.

Lovelock, J. E. (2006). *The revenge of Gaia: Earth's climate crisis & the fate of humanity*. New York: Basic Books.

Lovelock, J. E. (1979). *Gaia: A new look at life on earth*. Oxford: Oxford University Press.

Lovelock, J. E., & Margulis, L. (1974). Atmospheric homeostasis by and for the biosphere: The gaia hypothesis. *Tellus, 26*, 2-9.

Reilly, M. (1962). Occupational therapy can be one of the great ideas of 20th century medicine. *American Journal of Occupational Therapy, 16*, 1-9.

Table of Contents

Introduction

A friend of mine reviewed the first draft of the manuscript for this book and made a very insightful comment. She used the metaphor of a man holding a machine gun spraying bullets over a wide range but not really hitting the target (stated purpose of the book). The purpose, as she correctly pointed out, was based on the premise that human occupational behavior has global consequences that affect the entire planetary system. Subsequently, it was postulated that if individuals understood these far reaching repercussions of their occupational performance, they would be willing to change their behavior more readily in such a way that the planet and its ecological system were affected positively. Such understanding, it was argued, could be illuminated by knowledge derived from occupational science. The purpose of the book was to demonstrate how knowledge derived from occupational science could be used to inform individuals how they could change their occupational behavior for the benefit of the planetary ecological system. She did not think that this objective had been met.

After thinking about my friend's feedback, I realized that perhaps one of the major problems was that I had not explained the logic used to develop the book so that the reader could follow my thinking process. This introduction was added to explain that logic in order, hopefully, to demonstrate how all the chapters presented are related to each other and to the purpose of the book. I would like to begin by explaining that my thinking as I wrote this book was highly influenced by my education as an occupational therapist. Lovelock (2006) suggested that Gaia (his term for what might be understood as the planetary ecological system) is sick and he is the physician who is diagnosing and prescribing its treatment.

I would like to propose that if Lovelock is Gaia's physician, occupational scientists should be her therapists providing her with rehabilitative interventions. Their knowledge about human occupations derived from occupational science could be highly informed by occupational therapy practice among other factors. Occupational therapists are taught that the first step in dealing with a client is to identify his/her occupational performance issues (limitations in performance of self-maintenance, productivity, and leisure occupations that he/she wants, needs, or is expected to perform according to his/her age and cultural expectations). By identifying the occupational performance issues, a therapist clarifies the "what" of the problem. The second step in the rehabilitation process is to determine why that problem exists. If the person is not able to dress himself/herself, shop for groceries, go to work in order to earn a living, or play a game of golf that he/she used to enjoy, the therapist asks himself/herself: "why is this person unable to do these things?" May be due to an illness, injury, or age: the person is weak and does not have the endurance to stand long enough to accomplish a work task; has limited mobility which impends attempts to accomplish tasks; or has emotional, cognitive, or both emotional and cognitive impairments that limit his/her ability to sequence steps and problem-solve in order to complete required tasks.

In the third step, the therapist develops a *plan* of intervention to help the client overcome or compensate for identified limitations in order to perform occupations that he/she wants, needs, or is expected to do successfully. The plan consists of clearly stated functional goals and intervention activities and strategies to facilitate achievement of the goals. Finally, the client's performance is closely monitored to determine whether the desired outcomes are being realized.

Similar to the above described process, in the first part of this book, the global problems are identified (the "what" of Gaia's problems). I am aware of my friend's concern that devoting 3 chapters (chapters 1 to 3) to problem identification may be akin to using a machine gun to spray bullets over a wide range with the risk of missing the target (stated purpose of the book). I will use the same metaphor to explain my rationale for the necessity of those three chapters. When I was young, we used to go hunting for squirrels, partridges, deer, and other small game that were destroying crops on the farm. When we took the dogs to the woods to hunt, we had to flush out those animals so that we could shoot them (with bows and arrows since we did not have guns). Without flushing them out, we would not have known where to aim our hunting missiles. Furthermore, when we disturbed the brush, a variety of game that we did not even realize were there surfaced. Then we decided which ones to shoot and which ones to let go for the moment.

Using the above analogy, chapters 1 to 3 can be compared to going out into the woods to flush out the problems that confront our world today. Granted that the identified issues are not nearly exhaustive, but they are enough to help us understand the magnitude of what confronts us. Also, just like we knew that there were a variety of game lurking in the woods when we went hunting, and we knew generally what they might have been (squirrels, partridges, deer, etc.) even though we did not know beforehand the specifics of how many we would find of what species, similarly, many of the problems that confront us today are generally known. Everyone knows that there is poverty, corruption, climate change, etc. The "flushing" process refers to bringing the problems into sharper focus so that we can know where to aim our weapons (solution strategies).

Using the analogy of therapy introduced earlier, once the prevailing global problems have been brought into sharp focus in chapters 1 to 3, their origin in human occupational behavior (the why of the problems) is discussed in part II (chapters 4 and 5). Finally, in part III (chapters 6 and 7), a plan of intervention is presented. Of course as any occupational therapist knows, therapy is not complete until the intervention plan has been implemented and the outcome evaluated to determine whether or not it has been effective. However, implementation and outcome evaluation are beyond the scope of this book. Those are hopefully tasks for a future project. For now, I am content to highlight some of Gaia's issues of concern, propose the contribution of human occupational performance to the exacerbation of those issues, and suggest an occupation-based, individual-centered plan of intervention. My hope is that this work will provide a new approach to understanding the global problems in question and stimulate further conversation among those who are concerned.

Another friend who reviewed the first draft of the book manuscript pointed out that I was rather one-sided and seemed to have a bone to pick with those who lean towards conservatism politically. He correctly suggested that such a stance would alienate certain individuals making it difficult for the book to achieve its primary objective: appeal to all people to pull together for the purpose of healing Gaia by modifying their occupational performance behavior. In response to these comments, I reflected on my feelings and my perspective about the world, politics, economics, etc. I remembered the words of one of my favorite writers, Professor Ngugi wa Thing'o, who once stated that every writer is a writer in politics (wa Thiong'o, 1981). He argued that there was no such thing as non-partisanship in writing. When an author picked up a pen and put it on paper, he/she necessarily took a position, representing a certain view among many in the community of humans. I believe that this partisanship is not only true in political and popular literature but also in academic scholarship. As Kuhn (1973) suggested in his revolutionary proposition of the construct of paradigms, objectivity in scientific scholarship is in all reality a myth. Every scientist brings certain experiences, biases, a worldview, etc. to bear on his/her inquiry. This perspective colors the prism through which information is viewed, gathered, and interpreted, and therefore what conclusions are drawn. This necessitates that a writer be clear about his/her position in regard to the worldview so that he/she can try to maintain vigilance in order to minimize the effect of that view on his/her discourse in the subject at hand. In addition, a writer should honestly disclose his/her perspective of the world so that the reader can be clear about the conclusions to draw from the writing.

As I was thinking about my friend's feedback, I came across a statement by Barack Obama, the popular senator from Illinois, in his book, *The Audacity of Hope* (2006). In the book's prologue, Obama stated:

> I suspect that some readers may find my presentation of …to be insufficiently balanced. To this accusation, I stand guilty as charged. I am a Democrat, after all; my views on most topics correspond more closely to the editorial pages of the *New York Times* than those of the *Wall Street Journal*. I am angry about policies that consistently favor the wealthy and powerful over average Americans and insist that government has an important role in opening opportunity to all. I believe in evolution, scientific inquiry, and global warming; I believe in free speech, whether politically correct or politically incorrect… Furthermore, *I am a prisoner of my own biography: I can't help but view the American experience through the lens of a black man of mixed heritage, forever mindful of how generations of people who looked like me were subjugated and stigmatized, and the subtle and not so subtle ways that race and class continue to shape our lives.* (p. 10, emphasis mine)

The above statement seemed to resonate with my world view. I too realized that I was angry about policies that reduced certain people not only in the USA but even more so in the third world into paupers. I was angry about the fact that many children in the third world continued to die from curable diseases, malnutrition, and to have extremely limited opportunities and hope of enhancing the quality of their future lives; all this while billions of dollars were being spent every week to kill and maim people in places such as Iraq. I was angry about the fact that just a fraction of such resources (used to build and maintain war machinery) could transform the world and the lives of my poor relatives of whom I care very much, and this fact seemed to escape those who had assumed stewardship of world resources.

I too realized that I believed in evolution, scientific inquiry and global warming. I believed in not only free speech but also in compassion and the need to treat every human being in the world with dignity. Like Obama, I realized that I was a prisoner of my own biography, which consisted of my past experiences as a poor black child struggling to survive and maintain a hopeful vision of a dignified future while growing up in a former British colony where my parents were treated like children at best and a little better than animals at worst. I realized that my world view tended to be closely linked to those experiences.

I therefore tended to be rather impatient with the veneration of selfishness propagated as a virtue by big business corporations and the rich people whose interests they tended to serve. Not that I thought there was anything wrong with being wealthy or enterprising. However, I abhorred the arrogance expressed by some of our rich friends when they argued that their wealth was purely a result of their hard work. I would have been a little happier if they acknowledged that their riches had been accumulated on the backs of so many poor people, like my mother, who worked all their lives with their own hands for a pittance in return. Many of those poor people who built all the wealth in the world with their labor and blood died prematurely due to diseases and poor nutrition among other problems. If many of our rich friends were cognizant of this fact, they probably would have been more inspired to be empathic and more willing to assist those who were less fortunate members of our society, not only in the USA but all over the world.

While I never doubted that hard work was a virtue, I also believed that there should be policies to ensure that there was fair distribution of the wealth created making it beneficial to everyone in the world. I thought that the well-being of our world and of humanity depended on our ability to eventually achieve that measure of fairness. I believed that if everybody was willing to commit to attainment of such fairness, it would be possible even to eliminate poverty in the world altogether. Furthermore, I recognized that my upbringing in the cultural context of the Meru people of Kenya imbued me with certain attitudes and values about wealth, the environment, and responsibility towards the world and other people.

Traditionally, among the Meru people, individuals who were wealthy considered their riches to have been bestowed upon them by God (*Murungu*) for safe keeping and stewardship. Such wealth was therefore not theirs to squander but rather to use to take care of others who were in need. The Meru people believed that natural resources (land and its bounty) belonged to all people and could not be appropriated by individuals for their own use denying others access to them. Some of the resources like water catchment areas and natural forests were considered sacred and no one was allowed to live there or to destroy them. Even today, community elders are resisting the Government's attempts to settle people in such areas because they believe that people should not interfere with such sacred grounds (Gachanga, 2006). The community also believed, and still believes today, in a collective approach to solving problems, whether those problems are social, economic, or environmental (Kariuki & Place, 2005). All these values were instilled in me by my parents, aunts, and uncles, who were all very well grounded in the Meru traditions. They contributed towards crystallization of my specific world view.

Having recognized my perspective of the world as stated above [a process known as bracketing in qualitative research (Lincoln & Guba, 1985; Speziale & Carpenter, 2003)], I revised the manuscript with a view to minimizing the distorting effect that such a subjective frame of reference may have on the subject matter discussed in the book. I tried to present facts where warranted without blaming any class of people who may be opposed to my worldview. I kept in focus the fact that the objective of this work was to suggest a way of taking personal responsibility for global issues of concern and from this sense of responsibility, working together as members of the human community to rehabilitate Gaia so that she continued to support life. Finally, at the end of each chapter, I added exercises that readers could use to guide them to reflect further on the material discussed in the chapter. I believed that such exercises would help people to attain some depth of understanding of the issues of concern and their responsibility in meliorating those issues.

References

Gachanga, T. (2006). Kenya: "The land is ours" [Electronic version]. *AfricaFiles at Issue Ezine, 4*. Retrieved March 2007, from http://www.africafiles.org/atissueezine.asp?issue=issue4.

Kariuki, G., & Place, F. (2005). *Initiatives for rural development through collective action: The case of household participation in group activities in the Highlands of Central Kenya*. Capri Working Paper # 43, International Food Policy Institute, Washington, DC.

Kuhn, T. (1973). *The structure of scientific revolutions* (2nd ed.). Chicago: University of Chicago Press.

Lincoln, Y. S., & Guba, E. G. (1985). *Naturalistic inquiry*. Berverly Hills, CA: Sage.

Lovelock, J. E. (2006). *The revenge of Gaia: Earth's climate crisis & the fate of humanity*. New York: Basic Books.

Obama, B. (2006). *The audacity of hope: Thoughts on reclaiming the American dream*. New York: Crown Publishers.

Speziale, H. J., & Carpenter, D. R. (2003). *Qualitative research in nursing: Advancing the humanistic imperative* (3rd ed.). New York: Lippincott Williams & Wilkins.

wa Thiong'o, N. (1981). *Writers in politics*. London: Heinemann Educational Books.

PART I
THE AILING PLANET

In part I of this book (consisting of chapters 1 to 3), the problems confronting our planet (earth) are discussed. These include: diseases (many of them curable and/or preventable) that cause loss of life and much suffering; extreme poverty which is exacerbated by increasing material inequality between countries and between individuals within countries; corruption that renders social institutions inefficient and dysfunctional; wars that threaten human species and exacerbate poverty and disease problems; overpopulation; and environmental destruction that threatens all forms of life on the planet. The discussion in this part of the book will set the stage for an argument that all the problems enumerated above can be attributed to human choices and actions as they engage in daily occupations as individuals. Therefore, one effective way to meliorate them may be by facilitating change in the way individuals think, make choices, and act as they go about in their daily occupational lives.

Chapter 1
Disease, Poverty, Corruption, and War

In a newspaper article, Murunga (2006) wrote that we might be living in the best of times, in view of our recent history. He argued that at least in the last few decades, even though there had been much publicity about terrorism, there had actually been fewer terrorist events than in the 1960s and 1970s. We also have the cell phone and the internet which he emphatically stated that he would not give up for Armstrong's walk on the moon in the 1960s. This appraisal of the state of our existence today may be so. We may be living in the best of times. However, we are also living in one of the most troubled periods of our planet's history.

I am sure that in the past, others have said the same thing about other periods of history. However, if we examine closely some of the problems facing humanity and all life on earth today, we can easily understand why some people may feel a little uneasy about our state of the world. A few of those problems will be discussed in this chapter: human suffering and loss of life due to diseases; poverty; dysfunctional institutions; and wars.

Deadly Diseases That Threaten Humanity

Beginning the later part of the last century, deadly diseases such as the Acquired Immunodeficiency Syndrome (AIDS) and Ebola have arisen. These maladies are killing millions of people a year. By the year 2005, it was estimated that approximately 40.3 million people worldwide were living with AIDS (World Health Organization [WHO]) (Andrews, Skinner, & Zuma, 2006). Although the rate of infection worldwide had declined significantly since the 1980s, 64% (or over 3 million) of the new cases came out of Sub-Saharan Africa. Also, although mortality rates resulting from AIDS had declined due to introduction of effective retroviral drugs, the pandemic still continued to claim many lives in poor parts of the world. This was clearly evident in Sub-Saharan Africa where by the end of 2003, 15 million children under the age of 15 had been orphaned according to Andrews, Skinner, and Zuma, implying that for each of those children, both parents had died from AIDS. This means that millions of adults who were in their prime had died in a continent already devastated by poverty and other problems.

Other easily curable diseases that claim many lives in poor parts of the world include malaria, which kills about 1 million people a year (Attaran, Barnes, Bate, Binka, et al., 2006), diarrheal diseases which kill about 1.5 million children a year in Sub-Saharan Africa alone (Editorial, the Lancet, 2006), tuberculosis, and many others. One can therefore deduce that millions of adults and children, particularly from poor parts of the world such as Africa, lose their lives every year due to diseases, many of them preventable and/or curable.

Poverty and Material Disparity in the World

Very closely related to suffering due to death resulting from diseases is poverty. Clearly, there are glaring economic disparities, which result in saddening loss of life and suffering in parts of the world. Here are some mind-numbing statistics: According to the British Broadcasting Corporation (BBC) news (2006), in 1998, 1.2 billion people lived in absolute poverty. When we examine the distribution of wealth in the world, it is apparent that most of those people living below the poverty line as defined by the World Bank (living on less than US $1.00 per day per individual) live in South Asia, Sub-Saharan Africa, East Asia, Latin America, and 'others' in that order. This is reflected in the Gross Domestic Product per-capitas (in thousands of US dollars) which, according to the BBC (citing the World Bank) are as follows: South Asia, $440.00; Sub-Saharan Africa, $500.00; East Asia, $1000.00; Middle-East, $2060.00; Latin America, $3840.00, and the rich countries (United States, Britain, Switzerland, Denmark, Germany, Japan, etc.), $25730.00. As is evident in the above numbers, it seems that the few rich countries which host a small proportion of the close to 6.6 billion people in the world (United Nations Fund for Population Activities [UNFPA], 2007) have 3.28 times the GDP percapita incomes of all the poor countries of the world combined (25730/7840). This statistical reality is best illustrated in the following often cited summary of the status of our world. In the summary, it is postulated that if the world population were reduced to 100 people, the following would be their distribution:

> 57 Asians, 21 Europeans, 14 Americans from the North and the South, and 8 Africans
> 52 would be women, 48 would be men
> 70 would be people of color, 30 would be whites
> 89 would be heterosexuals, 11 would be homosexuals
> *6 would own almost 60% of the whole world's wealth, and all 6 would be from the United States*
> 80 would live in substandard housing
> 70 would not be able to read and write
> 50 would be undernourished
> 10 would have a disability
> 1 would be dying, 2 would be newborns
> 1 would have a computer...and only 1 will have enjoyed [sic] higher education (Kronenberg & Pollard, 2006, p. 618)

Further, while the GDP percapita in East Asia, North Africa, and the Middle-East rose by 6.4% between 1991 and 2000 with subsequent fall of extreme poverty from 29.6 to 14.9%, in sub-Saharan Africa, GDP percapita fell by 0.4% per year in that decade and extreme poverty rose from 47.4% to 49% of the population (International Bank for Reconstruction and Development/The World Bank [IBRD/WB], 2005). The percentage of people living in extreme poverty in the continent could be even higher today. In other words about half or more of the sub-Saharan African population today live in extreme poverty (on less than US$1.00 per individual per day) (BBC News, 2006; IBRD/WB, 2005). This makes people in the African continent the second poorest in the world after those in South Asia.

Furthermore, the gap in income between the rich and poor countries is steadily increasing, with a sharp rise in the last few decades. For example, in 1950, the difference in income between rich and poor countries was 35 to 1 (BBC News, 2006). In other words, if a worker in one of the rich countries earned US $35.00 per unit time, a person doing the same work in a poor country earned US $1.00 for the same unit time. In 1992, the difference was 72 to 1. In the last decade, the percapita incomes have actually fallen in at least 50 countries in the world.

The above material disparities are reflected in the differences in quality of life between the two sides of the poverty divide in the world. The manifestation of maladies such as diseases discussed earlier is felt more acutely in poor than in rich countries. As an illustration, since the first diagnosis of AIDS in the 1980s, HIV infection has ceased to be a death sentence in the rich countries. Antiretroviral drugs have been developed that enable individuals with the infection to live a more or less normal life for years. In the USA for instance, it is estimated that more than 1 million people are living with HIV (Andrews, Skinner, & Zuma, 2006). In that period of time, there have been about 500 thousand deaths in the country due to AIDS. In Africa on the other hand, the number of deaths resulting from the disease continue to escalate, accounting for the rising child mortality rates, decreased life expectancy (which is on average currently about 47 years [UNFPA, 2007]), and the rising number of orphaned children in the continent.

The gravity of the situation may be grasped if you consider, according to Andrews et al., that the majority of orphaned children in the world (8 out of every 10) live in Africa, and the situation is getting worse. That is why 90 % of the world's orphaned children live in Sub-Saharan Africa (Blake, n.d.). It is projected that by the year 2010, 18 million African children under the age of 18 years will be orphaned as a result of death of one or both parents due to AIDS. This situation is directly related to poverty within the continent as compared to the rich countries in the West. For example, by Western standards, the cost of anti-retroviral drugs that lengthen the lives of those infected with HIV and enable them to live a good quality of life has decreased significantly (to about US$150.00 per patient per year for the cheaper combinations of anti-retrovirals) (Van der Borght, de Wit, Janssens, van der Loeff, et al., 2006). If you add the cost of laboratory monitoring (about US $100.00) the cost is about US $250.00 per patient per year. This may be easily affordable by a person living in a Western rich country. However, for many people in poor countries such as in Africa where half of the population or more subsist on less than US $1.00 a day, such drugs are out of reach. This explains the prevalent mortality rates as a result of AIDS and subsequent problems such as inability of parents in households to work, death of parents leading to orphaned children, etc., leading to increased vulnerability of children. Such children are at risk of malnutrition, are unable to go to school, and therefore have poor future prospects. This implies a continued cycle of poverty in the future. Other treatable diseases that kill millions of people in poor countries include malaria, pneumonia, typhoid, etc. (Attaran et al., 2006; Blake, n.d.). Malaria alone kills about 1 million people in Africa every year.

Another easily demonstrable manifestation of the disparity is the prevalence of substandard living conditions in poor countries compared to the rich countries. This is reflected in the summary of the state of our world presented earlier where it was pointed out that out of every 100 people in the world, 80 live in substandard housing. Many people who are in abject poverty in the world live in the worst of those substandard dwellings, namely, the slums. These are defined by the United Nations (UN)-Habitat as dwellings satisfying one or more of the following conditions: a number of individuals live under one roof; there is lack of supply of safe water and sanitation; the individuals do not have security of ownership of the dwelling; the dwelling does not constitute a durable house; and the living space is inadequate (Warah, 2003). In such dwellings, the neighborhood is overcrowded, hundreds of people share a single toilet (if one exists), and there is a constant threat of eviction. Garbage, some of it toxic, is piled everywhere, and raw sewerage runs freely on the ground where children play and adults walk on it (USAID, 2006). The dwellings themselves are constructed of discarded timber, tin and iron sheets, and polythene paper.

It is instructive that the poor countries of the world are host to most of the individuals who live in such dwellings (a significant proportion of the 80% of the world population who live in substandard dwellings). According to Warah, in the year 2001, there were 554 million slum dwellers in Asia, accounting for 60% of the total population of such individuals in the world. Africa was second with 187 million people (20% of the world slum dwellers) living in slums. In the developed world (rich countries), there were only 54 million slum dwellers, accounting for 6% of this population in the world. In other words, out of all the people who live in the worst of the substandard housing structures (slums), the majority live in Asia, followed by Africa, and a small percentage in the developed world (Europe, USA, Canada, and Japan). It is therefore quite clear that disparities in economic status between the rich and poor countries are directly related to the number of people living in substandard dwellings known as slums. Living in such conditions heightens the problem of diseases discussed earlier. There is always a risk of water-born diseases such as cholera and typhoid in such neighborhoods, and infection with HIV/AIDS is rampant.

Material Disparity Within Countries

Material inequalities are not only increasing between countries but also among people within countries. Such inequalities are the realities in many of the African and other poor countries in the world. This was well expressed by Ivan Squire, an alumnus of the University of Richmond who visited South Africa on a mission and observed that the rich-versus-poor dichotomy was evident everywhere he looked (University of Richmond, 2006). As a result of this situation, in Kenya for example, 10% of the population depends on relief food for survival every year and half of the country's population is undernourished according to the Action Aid, a British Aid Agency (Kimani, 2006). Such disparities are evident in places like Nairobi, particularly in the sprawling slums such as the Mathari valley (see figure 1-1). Here, people live in plastic and iron structures with raw sewerage flowing freely on the surface [figure 1-2]), while on the other side of the street are visibly affluent neighborhoods like Muthaiga.

Figure 1-1.
A section of the sprawling Mathari Valley slum in Nairobi, Kenya.

Figure 1-2.
A close-up picture of one of the structures that serve as
accommodation for families living in the Mathari Valley slums
of Nairobi, Kenya. The little gulley at the bottom of the picture
is made by raw sewerage that flows freely on the surface.

Such gross disparities persist such that some people continue to live in
desperate conditions in spite of the overall increasing national wealth. In
Africa where 50% or more of the population live in extreme poverty as
defined earlier, there is evidence that more money is coming into the
continent and some people are becoming very rich. In recent years, Africa has
become increasingly important as a non-OPEC source of oil that fuels the
world economy (Palmeri & Morrison, 2006). As an illustration, in the last
year, African oil, mainly from Angola and Nigeria among other countries,
accounted for 30% of the world output of one of the major American oil
companies (Exxon Mobile). China is also becoming a major player in oil
exploration in Africa and indeed is one of the countries benefiting from
African oil. Companies from these countries are injecting billions of dollars
in Africa in exchange for oil exploration and extraction rights.

Such increasing national wealth is evident in countries like Kenya in which there has been a remarkably high economic growth. In 2006, Kenya experienced 6% growth in her economy, placing the country ahead of the pack among the least developed nations in the world (Nation Media Group, 2007). However, the UN report also indicated that it was ranked as one of the most un-equal nations in the world. The manifestation of these inequalities is seen in establishment of sprawling slums in places like Nairobi (see figures 1-1 and 1-2 above). Therefore, the question is, since some countries in Africa are experiencing unprecedented economic growth, but more and more people in the continent are becoming poorer every year, where is all the wealth going?

The above described situation is further underscored by the statistics from Brazil where in 2005, 10% of the population had 46.7% and the poorest 10% had only 0.5% of the country's income (Ituassu, 2005). In that country, public expenditure grew by 43.4% between 1980 and 2000. Therefore, the government was spending money on public projects even though it seems that only 10% of the population benefited significantly from such expenditure. This is an indication that the distribution of resources in that country was not designed to reduce economic inequalities. Increasing material inequalities are not unique to Brazil or African countries. Studies indicate that the gap between the richest and poorest individuals in many countries has increased over the last few decades (Fuentes, 2005) [there are of course other studies that indicate a decreasing gap between the rich and poor between 1970 and 2000 (Sala-I-Martin, 2002). However, these are in the minority].

According to the World Bank, the rise in inequality has been particularly marked within nations (Fuentes, 2005). While there has been economic growth in 51 countries between 1980 and 2001, only in 30 of them was there a reduction in income inequalities among citizens. As reported by Merrill Lynch and Gemini in their annual World Wealth Report (The CNN Wire, 2007), wealth is becoming more and more concentrated among the few ultra-wealthy individuals (those with more than $1 million in investable assets, whose wealth totaled $37.2 trillion in 2006) worldwide. This is evident even in the United States where the issue of the widening gap between the rich and the poor has started gaining attention in the past few years (Knox, 2006). These inequalities have consequences not only in individuals' lives but also on economic growth. According to Fuentes, extreme inequalities have been shown to affect economic growth negatively.

Dysfunctional Institutions

One may argue that many of the problems identified above, and others that will be discussed later, can be attributed to failure of social institutions to fulfill the mandate for which they were created; to serve all citizens fairly, without prejudice. These institutions include the Legislature, Judiciary, Executive, economic facilities, religious bodies, etc. Failure of such institutions can be caused by many factors. However, there seems to be a consensus of opinion that the biggest threat to institutional functioning is inefficiency which is often a manifestation of corruption (Genaux, 2004; Githongo, 2005; Meon & Weill, 2006). There are also those who argue that in inefficient systems, corruption gets things done by "greasing" the wheels of institutional machinery (Meon & Weill, 2006). However, even if it was true that such benefits of corruption exist, they may not be justified considering the exorbitant cost and the demoralization that the practice causes in society (Githongo, 2005; Posner, 2005). Therefore, in this section, the problem of institutional corruption will be discussed in detail.

Before examining how prevalent corruption might be, it is useful to define it. It has been observed in literature that the concept 'corruption' is rather vague and ambiguous (Posner, 2005). An activity considered appropriate and even admirable in one set of circumstances, such as giving a waiter a tip (gift) in order to get better service at a restaurant can be characterized as corruption in another set of circumstances, such as giving a gift to a public official in order to get better service in an office. It has also been suggested that what constitutes corruption is difficulty to establish objectively because it occurs privately between individuals willingly involved in the practice and therefore it is a 'victimless crime' (Issafrica, 2001). Given all the ambiguities, what is corruption?

Genaux (2004) defined what she called 'political corruption' as an activity which included bribery of individuals in positions of power to influence them so that they gave preferential treatment to a person or a group of people so that this person or persons had unfair advantage over others in society. Corruption in this sense included nepotism (preferential treatment by a public official of relatives, friends, members of the same race or ethnicity, while executing his/her official duties) and cronyism (preferential treatment by a public official of his/her friends, business affiliates, etc. while executing of his/her official duties). She identified the following categories of corruption: physical destruction of property; moral deterioration; and

33

pervasion of institutions, customs, etc. to support individual or group self-interest (for example using church teachings by politicians to influence people emotionally so that they vote for them to enable them to assume official positions so that they can protect their own selfish economic interests).

According to Genaux, it could be argued that corruption referred to the seriousness of acts in handling state affairs or reforming state institutions. In other words, corruption was the term used to denote "any disruption noxious to the smooth running of the state" (p. 15), or "a decomposition of the body politic through moral decay" (p. 15). For example, "an official who secretly accepts a bribe to decide a policy issue differently than he otherwise would have" is corrupt (p. 16). The scandals of Abramoff and the issue of congress ethics vis-à-vis the lobbyists in the early 2000s come to mind as examples of this type of corruption in the USA. Genaux further cited political scientists Heidenheimer and Friedrich in her attempt to define the concept 'corruption' even further. According to her, Friedrich saw corruption as deviant behavior consisting of activities performed by individuals for personal gain at the expense of the public. This behavior constituted perversion, adulteration, and debasement of a public office.

Further, Genaux traced the origin of the word corruption to French political literature. She quoted statements illustrating how corruption was perceived to lead to pervasion of public institutions, such as the judiciary, with wide ranging deleterious effects. The cited text stated: "…greed which resides in the heart of the judges can provoke much evil" (pp. 18-19). This quote could be applied to any institution, not just the judiciary. In the end, corruption may be seen as arising from greed and it leads to injustice and all the consequences that may be expected to arise from such injustice.

The above definition of corruption as an activity involving bribing public officials to influence official acts so that they favor certain individuals, or use of a public office for personal gain, was supported by Githongo (2005). He defined it as "the abuse of public office for private gain" (p. 1 of 3). He saw it as falling into one of three categories comparable to those advanced by Genaux: petty corruption, grand corruption, and looting. Petty corruption involved junior public officials and the amounts of money changing hands were small. This happened for example when a motorist bribed a police officer so that a minor traffic infraction could be overlooked. In grand corruption, businesses and high ranking government officials were involved. Large sums of money were involved and the corrupt schemes encompassed big government contracts.

Finally, according to Githongo, in looting, huge amounts of money were involved such that collapse of institutions such as banks could be the ultimate result. He stated that corruption "…often involves, for example, the printing of currency to fund fictitious projects, using public revenues to award enormous contracts to individuals who never supply the goods or the services" (p. 1 of 3). This was the kind of looting that happened in Kenya in 1992 when the incumbent government was facing the risk of being ousted in the first multi-party democratic elections since the country's independence. The then Kanu government printed large amounts of money and cut deals involving billions of Kenya Shillings with corrupt business people in order to raise revenue to buy voters. Kanu won the elections and the government retained political power. However, soon after the elections, inflation increased by over 100% and the economy was on the brink of collapse.

Incidences of corruption as defined above are numerous in the world, indicating significant dysfunction of social institutions. In the USA, there may not be petty corruption as defined by Githongo but there is evidence that there are incidences of grand corruption. Indicators of such corruption include the Enron scandal, the ethical issues regarding corruption of legislators by lobbyists in order to influence legislation (as in the Jack Abramoff case), etc. Ruskin (2006) cites many corrupt practices involving US legislators such as congressmen staffers accepting gifts so as to do favors for donors, allegations of converting campaign funds into foreign and corporate contributions, and an inefficient House Ethics Committee which is not willing to discipline house members who are involved in questionable practices. The above practices have led to a distrust of the politicians in general by the public. Fournier (2007) reports that many Americans are "dispirited" about their leaders, with 78% of the US citizens in a CNN opinion poll expressing their perception that the government is broken. According to the poll, many Americans believe that there is a leadership crisis in the USA.

Similar incidences of dysfunction and subsequent distrust of the government by citizens are evident in many other countries, especially in the third world. As an example, in Kenya the political class has been referred to as a gang of thieves who have taken the nation hostage (Sunday Nation Editorial, 2006). This perception was prompted by the fact that: "The list of senior politicians and current or former civil servants who were in service during the contracting of fishy projects is an unprecedented admission by the government of the existence - past and present - of widespread fraud and high level corruption" (p. 1 of 3). According to Transparency International (TI), a body that monitors the vice throughout the world, such corruption costs Kenya KSh 85 billion a year, an amount of money that would be enough to fund free primary education for at least 10 years, or to eliminate extreme poverty within the country altogether (Namunane, 2006). The country is ranked among the most corrupt nations in the world. Other dangerously corrupt nations according to the TI report, most of them in the African continent are Uganda, Tanzania, Zimbabwe, Nigeria, Pakistan, Sierra Leone, Cote d'Ivore, Democratic Republic of Congo, Sudan, Iraq, and Haiti. According to the report, the level of corruption in the USA is increasing significantly as well.

Other countries in which corruption is getting worse according to the report are Cuba, Brazil, Israel, Tunisia, and the Seychelles. It has been observed that corruption is particularly a problem in the poor countries of the third world where it is now well recognized "as a key factor in the effectiveness of some of their programs in various parts of the third world" (Githongo, 2005, p. 1 of 3). In other words, corruption is now recognized as a major contributor to dysfunction and failure of institutions in the third world countries that make elimination of poverty and economic inequalities difficult.

Institutional dysfunction and failure that leads to general despair among many individuals in this planet is not limited to government leadership. The Church has also been implicated in improprieties that are indicative of a possible source of distrust of the institution. The problem is not only the recent scandals in the Catholic church where the clergy were implicated in sexual abuse of children; or even the more recent scandals in the US involving politically right wing conservative politicians such as Mike Foley (whose political base was "Christian-based values") but also an even more basic departure of what might be seen as the religious (especially Christian) mission in society[1].

It seems quite evident that Jesus Christ saw the Church as a haven for poor people who may otherwise have no hope in the world. Thus, in his teachings, he commanded the church to care for the poor. This command was illustrated in the following parable (Jesus tended to teach through parables many times) about final judgment based on his followers "caring" for him or doing his will: "And the King answering shall say to them: Amen I say to you, as long as you did it to one of these my least brethren, you did it to me" (Mathew 25:40). Following are a few other quotes illustrating Jesus' concern for the poor: "For I was hungry and you gave me not to eat: I was thirsty and you gave me not to drink" (Mathew, 25:42). I was "sick and in prison and you did not visit me" (Mathew 25:43).

There is no doubt that according to the above cited verses, Jesus was very clear that it was the business of the Church and Christians to protect the poor and disenfranchised and to promote justice. It therefore stands to reason that the Church and Christians should be at the forefront of speaking loudly against practices that exacerbate existence of poverty, especially extreme poverty as defined by the World Bank, where individuals live on less than one dollar a day (and there are about a billion such people, most of them in Africa). Pope John Paul II articulated this mission of the church very well when he emphatically stated the following in his address to slum dwellers in the Philippines:

> Do not say that it is God's will that you remain in a condition of poverty, disease, unhealthy housing, that is contrary in many ways to your dignity as human persons. Do not say, "It is God who will it (sic)". (Second Plenary Council of the Philippines [PCPII], 1991, para. 1)

The teachings of the second Vatican Council also underscore Pope John Paul II's sentiments as indicated by their statement that "every person has the right to possess a sufficient amount of the earth's goods for himself and his family" (GS 68, cited in para. 1).

The problem is that there seems to have been an attempt to re-interpret the scriptures by some Church leaders and scholars in an attempt to abscond the responsibility of speaking up against injustice and standing up for the poor. In this re-interpretation, it is argued that when Jesus Christ talked about the poor being blessed, he did not mean that the church had an obligation to advocate for those who live in material poverty. He meant spiritual poverty. In this re-interpretation, a rich person could be even poorer in spirit than a materially poor person and therefore the Lord's favorite because of his/her humility. This was best stated by Keyes and Gallagher (1959) thus: "The rich often prove that they are not only capable of, but on occasion display, becoming humility. The have-not's, considered better schooled in the ways of humility, nonetheless possess the ability at times to be arrogant" (p. 209). I am not sure what poverty of spirit among the rich or arrogance of the poor means.

It may be argued that the attempt to subvert Jesus' teachings is responsible for the neglect of the poor by the Church and by some Christians. That is why those who are most vocal about religious values seem to be also the most critical of the poor and disadvantaged and the most likely to blame them for their circumstances. In one study, Hunt (2002) found that adherents of the Catholic Church were the most likely to believe that people were poor because of personal factors such as lack of ability, wastefulness, inability to save money, laziness, etc. (a perspective commonly known by social scientists as individualistic attribution of poverty). Catholics were followed by Protestants and Jews in that order in subscription to that belief. This was not the first time that studies indicated a tendency for Christians to blame the poor for their circumstances. Kluegel and Smith (1986) similarly found that individualistic attributions of poverty were most common among Protestants and Catholics.

This tendency to blame the poor for their poverty may be the reason that Church leaders are not so vocal in their advocacy for the disadvantaged poor. Rather, their concern seems to be homosexuality and abortion more than any other issue (see for example the sermon by Stedman, 1989). I am not saying that they should not preach against homosexuality and abortion if those practices are against their values. However, it seems hypocritical for Christian leaders to ignore and even seem to condone issues such as poverty, gross inequalities in society, exploitation of the poor and disadvantaged, etc. These are the issues that really cause suffering in the world. According to Baptiste and Damico (2005), the question of class is resurfacing in the USA because of increasing polarization between the rich and the poor. Poor people are unable to obtain medical care because they do not have insurance, are working longer hours with less 'real' wages, are constantly losing jobs as companies downsize and outsource, etc. These conditions drive more and more middle-income people into poverty.

Yet, to my knowledge, very few Christian leaders and preachers have spoken out aggressively against such trends which cause so much suffering to millions of people not only in the USA but all around the world. This is perhaps why many poor people are disillusioned with the Church. This disillusionment was best expressed by Blaine (2006), a Jamaican journalist who described her experience in a church in Jacksonville, South Carolina where she was witnessing the launch of "The New Church" by a pastor Daryl O'Neal. The pastor suggested that people were leaving the church because: "They are sick and tired of religion. They want God. They don't want the Church!" and the Church is "out of touch, out of reach, and out of action with the basic needs of people to be loved and to be shepherded" (Para. 7-8).

Blaine continued to observe that "no wonder the majority of the poor are disenfranchised and feel unwanted within the white-washed church walls" (para. 10). She further observed: "Almost every time I meet a Christian these days and they are fired-up about Christ, about the poor, and about God's purpose for Jamaica, they are either outside the Church or they are on the verge of leaving" (Para. 2).

The above statement indicates that many people perceive the Church as not following Christ's teachings to care for the poor, and not being in touch with God's purpose. Therefore, Blaine posed the following rhetorical question: If the Church will not champion the cause of the poor and advocate and defend them, who will? Her primary argument was that the Church was part of the reason that the poor were largely abandoned throughout the world. As observed in an earlier footnote however, I do not mean to suggest that all Christians are unfeeling and out of touch with the Church's mandate to serve and protect the poor. There are individual Christians who selflessly dedicate their lives to the service of the poor and to peace initiative so as to improve the lot of humankind. A very good example is Fr. Emmanuel Katongole, Associate Research Professor of Theology and World Christianity and Co-Director of the Center for Reconciliation at Duke Divinity School. Fr. Katongole has worked tirelessly for reconciliation in troubled spots in Africa and for initiatives to uplift the poor lot in the continent.

In another example, it was recently reported from Kenya that the various Christian Church leaders have joined with the Civil Society groups to demand that members of parliament and the president stop selfishly awarding themselves huge salaries and use that money to help the poor in the country. Another example is the Rev. Mutava Musyimi, Chairman of the Kenya National Anti-Corruption Steering Committee, who has been very vocal in calling for prosecution of those implicated in corruption cases such as the Anglo Leasing and Goldenberg scandals (discussed later) in which millions of Kenya Shillings that could have improved the lot of the poor were lost (Opiyo, 2007). There are many Christians like Fr. Katongole and the reverend Mutava Musyimi who are sincere in their service, seeking to find solutions to problems affecting the most disenfranchised in the world. However, the argument here is that the Church as a worldwide institution, for whatever reason, seems to have been subverted and its mission is no longer clear, at least not to the poor and the suffering.

Wars

Closely related to dysfunctional institutions are wars that cause loss of life and suffering and pose a threat to all humanity. We are living at a time in history when there are conflicts prompting the kind of war that has never been seen in the world before. This type of war (dubbed the 'war on terror') is unique because there is no end in sight, and political solutions seem to elude humanity (Bush, 2006a/2001, 2006b; The Economist, 2006; Meacham, 2006). These conflicts are mostly located in the Middle East but they are constantly threatening to overflow to the rest of the world.

The impact of such wars is evident in the number of lives lost. It is estimated that since the US led invasion of Iraq, about 655000 Iraqis have died as a result of war related violence (Burnham, Lafta, Shannon, & Roberts, 2006). This is in addition to those who died during the first Gulf war and many others who have died as a result of the six Arab-Israeli wars (Curtiss, 2006). The Arab-Israeli conflict continues as indicated by newspaper headlines such as the following: "Israeli tank shells kill at least 18 Palestinians" (Associated Press, 2006, p. 1 of 3). The devastation caused by these wars includes destruction of infrastructure and other material property, leaving thousands, perhaps millions of people in destitution. Images of such physical destruction in Iraq frequently confront us on television sets.

In addition to the Middle Eastern problem many people may be familiar with the genocide going on in Darfur, Sudan, where the Islamic government is sanctioning slaughter, torture, and rape of black Africans in the West. It will be pointed out in part II of the book that these wars can be attributed to human decision making while in pursuit of daily occupations. It will also be pointed out that many of these wars have an historical context and those who are responsible of decision making need to understand those contexts in order to make responsible decisions that will help end the wars and related human loss of life and suffering.

Reflection Exercise #1

After reading chapter one, think about the issues discussed in the chapter and answer honestly and as accurately as possible the following questions:

1. Before reading this chapter, how aware were you (on a scale from one to ten, with one being no awareness and ten being completely aware) about the following problems in the world?

   a.   There have been over 30 new diseases that have been diagnosed since the 1970s
   b.   Diseases such as malaria are becoming increasingly more prevalent and fatal as the planet becomes warmer
   c.   A large proportion of people in the world live on less than 1 US dollar a day
   d.   About 1 billion people in the world live in substandard accommodation structures commonly known as slums
   e.   About 655000 Iraqis have died as a result of the war that started in 2003
   f.   Over 200000 people have been killed in Darfur, Western Sudan by the Karhatoum government-supported Janjaweed militia
   g.   Many of the above problems can be attributed to failure of social and government institutions due to corruption and inefficiency

2. How can you get yourself further informed about the above issues?

3. What can you do as an individual to contribute towards making any of the above issues better?

References

Andrews, G., Skinner, D., & Zuma, K. (2006). Epidemiology of health and vulnerability among children orphaned and made vulnerable by HIV/AIDS in sub-Saharan Africa. *AIDS Care*, 18, 269-276.

Associated Press. (2006, November 8). Israeli tank shells kill at least 18 Palestinians [Electronic version]. *USA Today*. Retrieved November 8, 2006, from http://www.usatoday.com/news/world/2006-11-08-israel-Palestinians_x.htm?csp=24.

Attaran, A., Barnes, K., Bate, R., Binka, F., d'Alessandro, U., Fanello, C. I., et al. (2006). The World Bank: False financial and statistical accounts and medical malpractice in malaria treatment. *The Lancet*, 368, 247-252.

Baptiste, W., & Damico, N. (2005). Building the new freedom Church of the poor. Retrieved November 5, 2006 from http://www.findarticles.com/p/articles/mi_m2096/is_3_55/ai_n15950638.

BBC News. (2006). Attacking world poverty. Retrieved October 28, 2006 from http://news.bbc.co.uk/1/hi/business/924640.stm.

Blaine, B. A. (2006). The new Church and the poor. *Jamaica Observer*. Retrieved from http://www.jamaicaobserver.com/colums/html/20060703-180000-0500_108331_OBS_THE_NEW_CHURCH_AND_THE_POOR.asp.

Blake, C. E. (n.d.). Improving the lives of the most vulnerable children in Africa and the diaspora: Programs that work. Retrieved October 29, 2006 from http://www.saveafricaschildren.com/site/pageserver?pagename=programs.

Van der Borght, S., de Wit, T. F., Janssens, V., van der Loeff, M. F., Rijckborst, H., & Lange, J. M. (2006). HAART for the HIV-infected employees of larger companies in Africa. *The Lancet*, 368, 547-550.

Burnham, G., Lafta, R., Doocy, S., & Roberts, L. (2006). Mortality after the 2003 invasion of Iraq: A cross-sectional cluster sample survey. *The Lancet*, 368, 1421-1428.

Bush, G. W. (2006, October). Let's roll: President discusses war on terrorism. *Vital Speeches of the Day, 72*(24), 699-701. Speech first delivered to the Nation World Congress Center, Atlanta, Georgia, November 8, 2001.

Bush, G. W. (2006, September 5). Global war on terror: National strategy for combating terrorism. *Vital Speeches of the Day, 72*(24), 666-671.

Curtiss, R. H. (2006, November). In sixth Arab-Israeli war, Hezbollah survives, Israel loses, Bush missing in action. *The Washington Report on Middle East Affairs, 25*(8), 12-13.

Editorial. (2006a). Water and sanitation: The neglected health MDG. *The Lancet*, 368, 1212.

Editorial. (2006b). Poverty reduction needs both aid and trade. *The Lancet*, 368, 706.

Fournier, R. (2007 March 8). Public faith in leaders may be ebbing [Electronic version]. *USA Today*. Retrieved March 8 2007, from http://www.usatoday.com/news/washington/2007-03-08-leadership_N.htm?csp=24.

Fuentes, R. (2005). *Human development report 2005 - Human development report office occasional paper: Poverty, pro-poor growth and simulated inequality reduction*. No Place: United Nations Development Program.

Genaux, M. V. (2004). Social sciences and the evolving concept of corruption. *Social Law & Social Change*, 42, 13-24.

Githongo, J. (2005). Corruption as a problem in the developing world: Effects on the economy and morale presentation to the seminar on corruption and development co-operation held by the Government of Finland in May 2000 John Githongo, member of the Board of Transparency International, executive director of TI Kenya. *Key TI Speeches*. Retrieved October 31, 2006 from http://legacy.transparency.org/speeches/githongo/html.

Hunt, M. O. (2002). Religion, race/ethnicity, and beliefs about poverty. *Social Science Quarterly*, 83, 810-831.

International Bank for Reconstruction and Development/The World Bank. (2005). *Pro-poor growth in the 1990s - Lessons and insights from 14 countries: Operationalizing pro-poor growth research program*. Washington DC: Author.

Issafrica.org. (2001, September). Extent, locations and seriousness of corruption. *Monograph No 65: Corruption in South Africa, Results of an expert panel survey* (Chapter 3). Retrieved October 31, 2006 from http://www.issafrica.org/Pubs/Monographs/No65/Chap3.html.

Ituassu, A. (2005, October). In Brazil, inequality is a heirloom passed from one generation to the next [Electronic version]. *Brazil Magazine, 15*, 34. Retrieved October 31, 2006 from http://www.brazzil.com.

Keyes, N. B., & Gallagher, E. F. (1959). *Hope of the nation*. Gastonia, NC: Good Will Publishers.

Kimani, P. (2006, October). Writing key test on honesty [Electronic version]. *Daily Nation Online*. Retrieved October 29, 2006 from http://www.nationmedia.com/dailynation/nmgcontententry.asp?category_id=39&newsid=84457.

Kluegel, J. R., & Smith, E. R. (1986). Beliefs about inequality. New York: Aldine De Gruyter.

Knox, N. (2006, November 24). Wealth gap swallows up American dream. *USA Today*. Retrieved November 24, 2006, from http://www.usatoday.com/money/perfi/housing/2006-11-24-luxary-homes-usat_x.htm?csp=26.

Kronenberg, F., & Pollard, N. (2006). Political dimensions of occupation and the roles of occupational therapy. *American Journal of Occupational Therapy*, 60, 617-625.

Law, M., Polatajko, H., Baptiste, S., & Townsend, E. (2002). Core concepts of occupational therapy. In E, Townsend (Ed.), Enabling occupation: An occupational therapy perspective (pp. 29-56). Ottawa, ON: Canadian Association of Occupational Therapists.

Law, M., Baptiste, S., Carswell, A., McColl, M. A., Polatajko, H., & Pollock, N. (1998). *Canadian Occupational Performance Measure (3rd ed.)*. Ottawa, ON: CAOT Publications, ACE.

Meacham, J. (2006, November 6). The editor's desk: Top of the week [Electronic version]. *The Newsweek*, 4. Retrieved November 8, 2006 from http://proquest.umi.com/pqdweb?index=0&sid=1&srchmode=1&vinst=PROD&fmt=3&st.

Meon, P., & Weill, L. (2006). Is corruption an efficient grease? A cross-country aggregate analysis. PCS - JEL Classification: C33, K4, 011, 047, 1-31.

Mugonyi, D., & Barassa, L. (2007 March 19). Surrender half your pay, MPs told [Electronic version]. *Daily Nation Online*. Retrieved March 19, 2007, from http://www.nationmedia.com/dailynation/nmgcontententry.asp?category_id=1&newsid=94007.

Murunga, A. (2006, October). We could be in the best of times [Electronic version]. *The Daily Nation*. Retrieved October 28, 2006 from http://www.nationmedia.com/dailynation/nmgcontententry.asp?category_id=25&newsid=84332.

Namunane, B. (2006, November 6). Revealed: Graft costs Kenya KSh85bn per year. *Daily Nation Online*. Retrieved November 6, 2006 from http://www.nationmedia.com/dailynation/nmgcontententry.asp?category_id=1&newsid=85039.

Nation Media Group. (2007, July 20). UN report ranks Kenya ahead of World poor nations [Electronic version]. *Daily Nation Online*. Retrieved July 21, 2007, from http://www.nationmedia.com/dailynation/nmgcontententry.asp?categoryid.

Opiyo, D. (2007, June 28). List of shame on graft culprits proposed [Electronic version]. *Daily Nation Online*. Retrieved June 27, 2007, from http://www.nationmedia.com/dailynation/nmgcontententry.asp?category_id =1&newsid=101232.

Palmeri, C., & Morrison, M. (2006, October). Striking it rich in Africa [Electronic version]. *Business Week Online*. Retrieved October 31, 2006 from http://www.businessweek.com/magazine/content06_33/ b3997056.htm?campaign_id=rss_magzn.

Posner, B. (2005, August 28). Economics of corruption. Retrieved October 30, 2006 from http://www.becker-posner-blog.com/archives/2005/ 08/economics_of_co.html.

Ruskin, G. (2006). The fight against congress has collapsed. *Commercial Alert*. Retrieved November 1, 2006 from http://www.commercialalert.org/ issues/government/corruption/the-battle-against-corruption.

Sala-I-Martin, X. (2002). The disturbing "rise" of global income inequality. *Columbia University, New York, Discussion Paper # 0102-44*. Retrieved November 24, 2006, from http://www.columbia.edu/cu/ economics/discpapr/DP0102-44.pdf.

Second Plenary Council of the Philippines [PCPII]. (1991). The Church of the poor. Retrieved October 26, 2006 from http://aegis.ateneo.net/ jimeneeb/The%20Church%20of%20%the%20poor.htm.

Sunday Nation Editorial. (2006, September 24). Thieving political class has taken the nation hostage. *Sunday Nation Online*. Retrieved October 23, 2006.

Stedman, R. C. (1989). The poor-rich Church. Peninsula Bible Church Library. Retrieved November 5, 2006 from http://www.pbc.org/library/files/ html/4195.html.

United Nations Fund for Population Activities. (2007). State of world population in 2007: Unleashing the potential of urban growth [Electronic version]. Retrieved June 28, 2007, from http://www.unfpa.org/swp/2007/ english/introduction.html.

University of Richmond. (2006, June). Ivan Squire, '97, From missionary in South Africa to financial analyst at Ford Motor Company [Electronic version]. *Jepson: An Electronic Newsletter for Alumni and Friends of the Jepson School of Leadership*. Retrieved October 31, 2006 from http:// www.oncampus.richmond.edu/academics/leadership/newsletter/spring2006/ ivan_squire.htm.

Warah, R. (2003). Slums are the heartbeat of cities [Electronic version]. *Global Policy Forum*. Retrieved October 31, 2006 from http:// www.globalpolicy.org/socecon/develop/2003/1006slums.htm.

Chapter 2
Over-Population: A Real Problem or Simply a Scare from Pessimists?

In the year 1798, Thomas Robert Malthus, a political economist and Anglican Parson who hailed from Dorkin, South London, published a book entitled, *An Essay on the Principle of Population as it Affects the Future Improvement of Society*. In the essay, he sought to investigate and explain how human population growth was regulated by available natural resources. This intention was apparent in the preface of the book, where he wrote: "It is an obvious truth, which has been taken notice of by many writers, that population must always be kept down to the level of the means of subsistence; but no writer that the author recollects has inquired particularly into the means by which this level is effected" (p. 1). About two decades earlier, Adam Smith (1776), dubbed the father of Capitalism, had published his famous book, *The Wealth of Nations*, in which he explored how necessity and human selfish drives contribute to accumulation of wealth that benefits everybody in society.

In this work, Smith suggested that it was not human generosity and philanthropic desire that led to creation of wealth that sustained human populations. For example, the butcher did not work hard simply because he desired to feed other peoples' families, or for that matter, the car maker in order to help people travel more easily and comfortably. Rather, these individuals were motivated by self interest (to make wealth for their own use). His contention was that this selfish trait in human beings could be harnessed to generate wealth that benefited everybody in society. This harnessing of self-interest to benefit the public would be achieved by establishing a free market system where producers of material objects and services competed. In this system, human business endeavors that were creative and adaptive to human needs, and maximized production of things and services that were needed by the public with minimal input of resources survived. Those that were not able to compete died off. This competition led to continual improvement of products, both in quality and quantity, which resulted in creation of wealth that benefited all members of society.

It is reasonable to assume that Malthus was somewhat influenced by Adam Smith's work, as well as by other writers of his time. This influence was evident in his statement that his argument was by no means new. Rather: "The principles on which it depends have been explained in part by Hume, and more at large by *Dr. Adam Smith*" (p. 3, emphasis mine). However, Smith (1776) differed from Malthus in at least two ways: First, he did not seem to be concerned about the limitation of resources that our planet afforded and hence, the need to curb population growth to ensure that it remained below its means of subsistence. Although he seemed to be in agreement with Malthus that: "Every species of animals naturally multiplies in proportion to the means of their subsistence, and no species can ever multiply beyond it" (p. 47), to him, population growth was regulated by the same supply and demand laws that governed the rest of the economic world as he conceptualized it. Thus, he wrote that:

> ...the demand for men, like that for any other commodity, necessarily regulates the production of men; quickens it when it goes on too slowly, and stops it when it advances too fast. It is this demand which regulates and determines the state of propagation in all the different countries of the world, in North America, in Europe, and in China...(p. 48)

In other words, Smith's postulation was that when the supply of labor was too low, its price increased, making it possible for laborers to earn more than they needed to sustain themselves. They used the surplus to raise children. If the price of labor became too low, many of the laborers opted not to have children because they could not afford to provide for them. In that sense, it seems that Smith was of the opinion that the free market system was capable of effectively controlling population growth as it was in controlling the price and abundance of other commodities. Of course it is doubtful that falling wages are effective in reducing population growth. If that were the case, in many third world countries where labor is extremely cheap and unemployment rates unbelievably high, there would be negligible or even negative population growth.

The second way in which Smith differed from Malthus was that he saw population growth as a positive thing for the economy. This was evident in his statement that: "The most decisive mark of the prosperity of any country is the increase of the number of its inhabitants" (p. 41). To him, higher population meant not only a readily available labor which would lower the cost of production of things, but also a larger market where products could be sold for a profit. To illustrate this point, he gave the example of the price of cattle in a community with low population. He pointed out that in such a community, the price of the hide of the cow was much more profitable than the cow's meat, and often, a cow was killed for the hide and the carcass left to rot or to be devoured by vultures and other beasts of prey. Thus, he stated:

> In countries ill cultivated, and therefore but thinly inhabited, the price of the wool and the hide bears always a much greater proportion to that of the whole beast than in countries where, *improvement and population being further advanced, there is more demand for butcher's meat*. (p. 140, emphasis mine)

Malthus (1798) agreed with Smith that the population could not grow beyond the means of its subsistence. He postulated that there were two laws governing population growth: 1) Humans' need for food in order to exist; and 2) Their sexual passion. He saw the above as the two fixed laws of nature. In other words, human desire for food and sex could not be controlled. This was the basis of his disagreement with enlightenment writers such as Godwin and Condorcet who thought that human perfection could be achieved. In Malthus' view, since humans could not control their passions (for food and sex), human perfection was not achievable.

The problem was that satisfying the passion for sex led to population growth. Furthermore population growth tended to supersede the growth of means of subsistence. He hypothesized that the population grew at a geometrical ratio (1, 2, 4, 8, 16, 32, 64, 128, 256,), while subsistence (food) grew at an arithmetic ratio (1, 2, 3, 4, 5, 6, 7,). This meant that very soon, the population growth would overtake available resources to support it. He postulated that based on numbers available from the American colonies where population growth was unchecked, the number of people in the world would double every 25 years. If he were correct, based on the estimated world population which was about one billion people in 1830 when he was alive (Desip., 2006), today we would have 128 billion people in the world. He did not think that the earth was capable of supporting even a fraction of that many people.

Malthus further postulated that there were two ways by which nature curbed population growth so that it remained below the means of subsistence. These were misery and vice. When the number of people reached a point where available natural resources could not support them, misery and vice resulted. The two were expressed through famine, disease, and war. Through these means, many people were killed and the population was reduced to levels that could be sustained by available resources.

Therefore, Malthus theses, as articulated in his essay, could be summarized in the following three general propositions (Desip., 2006): Proposition 1 - Misery is the means by which the balance between population growth and the means of subsistence is achieved. During good times, people reproduce to the point where the means of their subsistence are overwhelmed. Misery is triggered in order to reduce the number of people and restore balance. Proposition 2 - The distress produced by misery due to overpopulation affects poor people the most. Proposition 3 - History and politics may be seen as a struggle for control of resources. This struggle is the basis of all conflicts that often lead to the fall of empires, such as the Roman Empire. In other words, overpopulation may be seen as the indirect cause of all conflicts, whereby humans struggle for control of resources that become scarce as the number of people increases.

As soon as Malthus published the essay, he was highly criticized, not so much for his arguments about population and its relationship to the means of subsistence, but more for what he wrote about how the poor should be treated in an attempt to control their reproduction and therefore to curb population growth. In the beginning, he proposed education of the poor into the luxuries of middle class as a means of encouraging them to limit their reproduction so as to attain that lifestyle. He also suggested that individuals be encouraged not to get married early. In later editions of his essay, he proposed a moral check to population growth, by encouraging the poor to abstain from sex voluntarily by not getting married. However, he also seemed to agree with Smith's notion of supply and demand as a means of controlling population growth as indicated by his advocacy in chapter five of his essay to abolish the poor laws that had been in effect in England for about 200 years, and which were meant to offer relief assistance to the poor. In his view, offering assistance to the poor relieved suffering only in the short-term. In the long-term, the suffering would be even worse, presumably because relief would encourage them to reproduce, increasing the number of people to a point where they could not be sustained with available resources.

Malthus' sentiments in this regard led to severe criticism of his ideas. Among other things, he was labeled a defender of small pox, slavery, and murder of children in order to keep the population low, and as being anti-soup kitchens, early marriage and assistance to the poor in an effort to discourage irresponsible population growth (Bonar, 1885; Hardin, 1998). Although Malthus' work was published over two Centuries ago, his propositions about the relationship between population growth and the means of subsistence afforded by the earth continues to be debated. In this debate, the neo-Malthusians have not been much help in popularizing his ideas. Their stance vis-à-vis the status of the poor and disenfranchised has made Malthusian ideas quite unpalatable. They not only advocate cutting welfare programs but some have gone as far as proposing that the rich Western countries should not assist starving nations because that would interfere with the natural checks to population growth. An example of this negative attitude towards the poor was the argument that food should not be wasted on those who cannot be saved. Those who held this view "went to the extent of devising a list of nations that should be allowed to starve" (Maher, 1995, p. 12).

Although many people may be alienated by Malthus' and neo-Malthusian sentiments and tendency to blame the poor for all population related problems, and to seemingly advocate their punishment, it may be beneficial to acknowledge their argument that we live in a non-elastic world. Space and natural resources are not increasing while human population is still growing. Surely, it is a matter of time before we reach a critical point, where like chickens in a coup, there are too many of us in this planet. Those who have raised chickens know what happens when there are too many of them enclosed in a limited space. They start fighting each other, and disease out-breaks wipe out a big proportion of them.

The same fate could face humans due to overpopulation in a limited space provided by the earth. It is even conceivable that it is already happening (which would validate Malthus' and neo-Malthusians' prediction that there would be outbreak of diseases, starvation, and war). In this chapter, this question will be explored by examining available evidence that: population growth is exerting pressure on the environment; unchecked population growth is resulting in scarcity of natural resources; there are wars over scarce natural resources occasioned by population growth; and there is an increase in emergence of new diseases which can be attributed to population growth.

Is Population Growth Exerting Too Much Pressure on the Environment?

It can be argued that population growth is resulting in climate change that may prove catastrophic over time. However, there have been disagreements over whether the climatic changes that we are experiencing are even significant. It is therefore difficulty to conclusively connect the change to overpopulation let alone make assertions about the relationship between the two variables and the effect of this relationship on the earth and life on it. This topic will be debated more exhaustively in the next chapter. In this section, I would like to point out that there is a documented relationship between population growth and environmental destruction, which may prove problematic for humans and other life in future. In the past two decades, human activities such as logging, increased farm acreage, etc., have led to deforestation of more than 120,000 square kilometers of land per year (Hunter, 2001; Meyerson, 2004). Re-aforestation activities have led to reclamation of only about one tenth of that area per year. This trend has resulted in destruction of more than half of the world's original forests.

While it may be difficulty to claim conclusively that population growth is a direct cause of this environmental destruction, it is important to note that between 1960 and 1999 (a span of about 40 years), the world population more than doubled from 3 billion to over 6 billion people (RAND, 2000). More specifically, in 2006, the world population was estimated to be about 6.6 billion people (United Nations Fund for Population Activities [UNFPA], 2007). In the same period, changes in environment, some of which have been mentioned above (deforestation, pollution, resource depletion, the threat of rising sea levels, etc.) accelerated as the population increased. It seems quite evident that with population growth, resources such as arable land, water, forests, and fisheries are being stretched to the limit.

Furthermore, according to the RAND, more and more people are migrating to cities. For the first time in history, more than half of the world population (3.3 billion people) is living in urban areas according to the UNFPA. By 2030, it is estimated that about 5 billion people (60% of the world population) will be living in cities. Most of this urbanization will be in developing nations. This trend is leading to more problems of pollution, poverty, shortage of resources such as clean drinking water, sanitation, etc.

Therefore, even though the relationship between population growth and environmental change is not easily directly verifiable, there is reason to believe that such a relationship is strong because of the link between population and deforestation, destruction of water sources, pollution, etc. I will illustrate this relationship with a personal example. I was born in a small new settlement known as Ruiri Sub-Location in Meru District, Kenya, on the Northern slopes of Mount Kenya. When I was a child in the 1960s I remember the place having many trees and all kinds of vegetation. I remember zebras, giraffes, and other animals grazing right on our backyard. There was a creek that ran through the sub-location and it passed less than a mile from my house. There were many other natural springs and related marshlands. However, all these natural sources of water were on land that had been allocated to individuals for cultivation. The owners of the land where the creek passed or where the springs were located did not prevent people from fetching water. According to the Meru traditions, they were expected to keep these water sources open to all members of the community. Moreover, I remember it raining almost all the time. Vegetation was green and lustrous.

The population in the entire area (about 121 square miles) consisted of about 100 families or less. Certainly, everyone knew everyone else in the entire location (which was then a sub-location). Each family had on average six or more children. The size of the land per family ranged between 10 and 30 acres (average, 15 acres). By the early 1970s, about a decade later, most of the trees and vegetation had been cleared to give way to farmland. People had cultivated along the banks of the creek and the natural springs, draining the marshlands and in the process drying up the springs. People living upstream had diverted water from the only creek using fallows for irrigation of their land. Eventually, the creek, which had run through the region for centuries, providing water to sustain thousands of animals, started drying up. I remember when I was in standard four (the equivalent of grade 4), my sisters and I were getting up at 3:00 am in the morning to go along the creek in search of puddles of water on the river bed from which we could draw water for drinking and cooking before going to school. Eventually, the creek dried up all together.

Around 1972, I witnessed the first draught which resulted in the first famine in my life-time. Since then, there have been frequent rain shortages in the area. Besides, whereas the number of families in my childhood was 100 or less (which meant a population of less than 1,000 people), all the children in those families grew up. The land was subdivided between those children who started raising their own families, consisting of between 2 and 8 children a family. As Barr, Tropical Forest Trust, and McGrew (2004) point out, when

Meru children are grown up, it is expected that their parents will divide the available piece of land between them. Therefore:

> When a young couple inherits land, usually from the husband's father, the piece of land is smaller than the father's holdings due to subdivision of land among the children. Delineating and establishing land boundaries, building a house, and preparing the land for cultivation are the top priorities. These tasks usually involve clearing trees in fields that will be used for crop cultivation and planting trees along the new boundary lines for delineation. It also involves planting fruit and fodder trees. (p. 44)

Subdivision of land as the population increases with coming of age of the children means clearing of indigenous trees even though some trees may be planted along the boundaries and fruit trees on a portion of the land. Also, many of these children did not go to college or acquire other skills. So they opted to stay on the land and try to eke a living out of it. Today, a family is lucky if they have more than 2 acres of land. Whereas the population in the area during my childhood was in the hundreds, today, there are at least 11,000 inhabitants in Ruiri location according to the 1999 census (World Bank, n.d.). The increased population meant that the land became over-farmed which led to impoverishment of the soil. Subsequently, many of the crops that used to do well on the land are not able to survive any more.

One can therefore deduce a clear correlation between increase in population between 1960 and 2006, and deterioration of the environment which included deforestation and soil impoverishment. In addition, there have been reports of increasing pollution of water sources in the area leading to people becoming ill (sometimes even dying) from use of water contaminated with raw human waste, pesticides, and hospital waste (Imathiu, 1998). In 1998, "...30 students from Ruiri Secondary School were treated at Meru Central District Hospital for diarrhea after drinking water that had been polluted with waste dumped in Mt Kenya Forest around Imenti" (p. 1 of 3). Ruiri is a perfect microcosmic example of environmental destruction as indicated by deforestation, pollution, and resource scarcity between 1960 and 1999 with the doubling (in the case of Ruiri increase by about 10 times) of the population (RAND, 2000).

Is There Scarcity of Natural Resources as a Result of Population Increase?

One of Malthus' (1798) postulations was that when population growth exceeded subsistence levels, natural checks consisting of misery and vice were activated. Indicators of those checks were famine (occasioned by scarcity), disease, and war. Based on this basic postulation, neo-Malthusians predicted that by 1970s, there would be famines resulting in millions of people starving to death (Maher, 1995). Other predictions were that: capital per worker in terms of buildings and equipment would decrease leading to a fall in work output percapita; increased population would lead to inability to educate people resulting in low average human capital; and population growth would lead to more consumption and less ability to save. Ehrlich (cited in Maher, 1995), one of the leading demographic scholars with neo-Malthusian leanings, predicted that there would be water rationing by 1974 and food rationing by 1980, and oceans could be dry by 1979.

These predictions have been the target of ridicule by anti-Malthusians. For example, O'Rourke (cited in Maher, 1995) pointed out that in the late 20th century, oceans had not yet dried up, and there was no water and food rationing. To O'Rourke (1995), there was no problem in the world today that could not be solved by good old capitalism, free market economy, and subsequent economic prosperity. Eberstadt (2006) suggested that neo-Malthusians had been consistently wrong in their predictions and all they managed to do was turn Americans among other people in the world into victims of quantophrenia. By quantophrenia he meant obsession "over numbers as descriptors, no matter how dubious their basis or questionable their provenance" (p. 28). Those duped by alarmist neo-Malthusians, according to Eberstadt, included:

> ...the United Nations, the World Bank, the U.S. Department of Agriculture, even the Central Intelligence Agency. Differing mainly in their presentation of details, the members of this grim chorus commonly asserted that the burgeoning number of mouths on the planet meant that more scarcity, poverty, and hunger were just around the corner - with the most severe suffering predicted for the rapidly reproducing Third World. (p. 28)

Disputing these predictions, Eberstadt argued that on the contrary: "Troubled as the world may be today, it is incontestably *less* poor, *less* unhealthy, and *less* hungry than it was 30 years ago" (p. 28, emphasis original). Many other criticisms of Malthusian ideas about the impact of over-population can be cited along those lines. As I mentioned earlier, I do not concur with Malthus and the neo-Malthusians on their tendency to blame the poor for over-population, and their proposed cruel measures to keep the population among the poor in check. Those disagreements notwithstanding, I think it would be the height of irresponsibility to ignore Malthus' warning and assume that the planet is capable of supporting a limitless number of people. Such an assumption defies logic. Given that the planet is not expanding in size or in the number of natural resources on it (in fact, these resources are constantly being exhausted, whether it is natural forests, minerals, etc.), at some point we are bound to reach the limit of growth where we will have too many people on earth.

It is true that technological advances led to increased food production and more efficient extraction of other resources. This development kept the means of subsistence ahead of population growth and resulted in the ability of the planet to support more people than Malthus had anticipated (Maher, 1995). "Hybrid grains, nitrogen fertilizers, mechanized cultivation, pesticides, irrigation - these all boosted crop yields beyond the wildest dreams of Malthus and his contemporaries" (Anonymous, 2004, p. 9). So, does it mean that Malthus was totally wrong? I think that technology simply postponed the inevitable. As has been observed, "There are disquieting signs that the run of success in keeping food production ahead of population growth is coming to an end" (pp. 9, 11). Examples of the above mentioned signs include the fact that:

> ...as global population continues to grow, limits on such global resources as land and water have come into sharper focus. For example, only in the later half of the twentieth century has the unavailability of land become a potentially limiting factor in global food production. (Hunter, 2001, p. xii)

In 2004, "the United Nations Agricultural Organization reported that for the fifth year running the global harvest had failed to produce enough food for everyone" (p. 11). Furthermore, there has been a decrease in cropland by about 20% per head of population in the last decade alone, not to mention that in the 28 poorest nations in the world, all of them in Africa, their citizens are nutritionally worse off than they were 30 years ago.

To give a concrete example, in the example of Ruiri given earlier, with deforestation and impoverishment of land through over-farming, many of the crops that used to do very well can no longer grow there. When I was a child, there used to be plenty of bananas, sweet potatoes, yams, arrow roots, sugar-cane, etc. It did not take much effort to grow these crops, and there was plenty to eat. Other crops that did well included black beans, snow peas, pigeon peas, sorghum, etc. Today, it is all one can do to raise a decent crop of corn and beans on the land. As a result, since the 1970s, I have seen several famines occur not only due to shortage of rain but also because nothing grows on the land that easily any more. I have also mentioned the water problems that resulted from drying up of the only creek and the natural springs that traversed the land. This scarcity was reflected in the fact that according to the 1999 census, 57% of people living in Ruiri (6,140/10,823) were below the poverty line (subsisted on less than US$1.00 per individual per day) (World Bank, n.d.).

Other resources that have become scarce with increasing population include wood fuel in places like Africa where wood is the only affordable form of fuel for most people, for heating and cooking (Hunter, Twine, & Johnson, 2005; Kirkland, Hunter, & Twine, 2005). This scarcity has led to further destruction of the environment through deforestation. As Kirkland et al. (2005b, p. 13) stated: "*dry wood are scarce and some people use electricity stoves but some don't, then they are forced to cut down living trees...*" (emphasis original).

The other resource whose scarcity is threatening to impact the world is oil. Kunstler (2005) asserted that: "We are in for a rough ride through uncharted territory" (p. 23) because cheap fossil fuels (oil and natural gas) are on the verge of exhaustion. In a world where food production and other products necessary to meet the needs of the burgeoning world population are dependent on oil, its shortage will have far reaching implications. Kunstler suggested that without fossil fuel, the world would degenerate to pre-industrial levels of functioning, and in those circumstances, the planet can only support one billion people, meaning that the remaining over 5 billion humans would die. In other words, those who criticize Malthusian predictions may have spoken too soon. We may yet be heading towards a crisis occasioned by uncontrolled population growth, unless something is done to correct the situation.

Are There Wars Occasioned by Population Growth?

The final indicator of misery and vice that would check population growth when it exceeded the resources available for its subsistence, according to Malthus, was war. Based on this prediction, neo-Malthusians claimed that not only were we going to see an increase in hatred, violence, war, and bloodshed as a consequence of scarcity due to increased population, but that all conflicts and politics are generally about control of resources (Anonymous, 2004; Desip., 1997, 2006; Landry, 2006). The question is, are there current wars that may evidently be "a struggle for control of resources that is triggered by overpopulation" (Anonymous, 2004, p. 9)?

Kunstler (2005) stated that: "The decline of fossil fuels is certain to ignite chronic strife between nations contesting the remaining supplies" (p. 23). He continued to assert that indeed wars over fossil fuels have already begun. In this regard, there are those who may argue that the Iraq war is about oil. Although it may be difficult to substantiate such a claim with concrete evidence, it is not out of the realm of possibility. A more direct illustration of war triggered by human competition for resources is the Darfur conflict, in which "After years of low-level clashes over *water* and *land* in the vast, arid Darfur region, rebels from ethnic African tribes took up arms against Sudan's Arab-dominated central government in 2003" (Associated Press, 2006, p. 2 of 3, emphasis mine). In that conflict, about 200,000 people have been killed so far. Another 2.5 million have been displaced from their homes.

It is also instructive that historically, such conflicts tend to occur during difficulty economic times (indicative of scarcity). The pogroms in Russia, inquisitions in Spain, and other incidences in Europe where thousands of Jews were killed occurred during such times (Tessler, 1994). Similar conflicts occur from time to time in countries such as Kenya, particularly in poor, highly populated areas such as the Mathari slums (Mugwang'a, 2006; Nation team, 2006). Even in the US, we are seeing strong rhetoric with racial overtones from both politicians and the general public against immigrants from the third world (particularly from Mexico) at a time when many middle and lower class workers are experiencing an economic squeeze. Similarly, we are seeing a resurgence of racially motivated statements and hate crimes in the country (Race Relations Reporter, 2006). Anti-immigration sentiments, directed primarily against Chinese immigrants, are increasing in Russia as well, according to a recent CNN report. Therefore, there is some evidence that as Malthus postulated, as the population growth approaches the limits imposed by available means of its subsistence, incidences of conflict over shrinking resources are on the rise.

Are There New Diseases That Can Be Attributed to Population Growth?

Finally, Malthus postulated that the last indicator of the natural checks to population growth (misery and vice) was disease. He argued that as the population growth reached the limits of its subsistence, diseases would arise that would wipe out a sizable proportion of people. Like all his other ideas, this postulation was similarly criticized. Eberstadt (2006) argued that rather than leading to disease infestation, population explosion has in fact been a 'health explosion'. This statement flies in the face of the fact that we are living in a time when the Acquired Immunodeficiency Syndrome (AIDS) is killing hundreds of thousands of people in poor countries in places such as Africa. Besides, according to Britain's Royal Society, 30 new diseases have been documented since 1975, AIDS among them, and 30 new diseases are expected to emerge in the next 30 years (Anonymous, 2004).

Furthermore, as will be discussed in the next chapter, there is evidence of a link between the activities of human beings and global warming, which is related to the devastating climate change that we are witnessing. This means that population growth is related to climate change because higher numbers of people mean more activities that cause pollution which is associated with global warming. Since warmer climate is associated with an increase in diseases such as malaria and yellow fever (Barasa, 2006), it follows that global warming due to increased population growth is responsible for an increase in such diseases.

Reflection Exercise #2

After reading chapter 2, answer the following questions:

1. Before reading chapter 2, how aware were you (on a scale of one to ten, with one being not at all aware and ten being completely aware) about the fact that:

   a. Between 1960 and 2006, there has been a growth of world population by more than twice from 3 to 6.6 billion people?
   b. The increase in population has been found to be correlated with environmental deterioration as evidenced by decreased forest cover, extinction of various animal and plant species, pollution of the atmosphere, etc.?

2. What can you do to become more informed about the relationship between overpopulation and various global problems that we face in the 21st century?

3. What can you do as an individual to make your contribution towards meliorating the problem of overpopulation and other associated problems?

References

Anonymous. (2004). Population - Thomas Malthus: Doctor doom. *Canada and the World Backgrounder, 70*(3), 7-12.

Associated Press. (2006, November 17). Agreement reached on Darfur peacekeepers. *USA Today*, Retrieved November 17, 2006, from http://www.usatoday.com/news/world/2006-11-17-un-darfur_x.htm?csp=24.

Barasa, L. (2006, November 16). Act or perish, warns UN chief. *Daily Nation Online*. Retrieved November 17, 2006, from http://www.nationmedia.com/dailynation/nmgcontententry.asp?category_id=1&newsid=85648.

Barr, R., Tropical Forest Trust, & McGrew, J. (2004). *Landscape-level tree management in Meru Central District, Kenya.* Agroforestry in Landscape Mosaics Working Paper Series. World Agroforestly Center, Tropical Resources Institute of Yale University, and the University of Georgia.

Bonar, J. (1885). *Malthus and his work.* London: Macmillan.

Cohen, E. (2006). Why have children? *Commentary: Research Library, 121*(6), 44-49.

Coren, M. (2006, February 10). The science debate behind climate change: Forecasting the future remains a contentious exercise. *CNN*. Retrieved November 20, 2006, from http://www.cnn.com/2005/TECH/Science/04/08/earth.Science/index.html.

Desip. (2006). *Malthus Society rationale and core principles.* Retrieved November 14, 2006, from http://desip.igc.org/malthus/principles.html.

Desip. (1997). *Predictions.* Retrieved November 14, 2006, from http://desip.igc.org/malthus/predictions.html.

Eberstadt, N. (2006). Doom and demography. *The Wilson Quarterly, 30*(1), 27-31.

Hardin, G. (1998, Spring). The feast of Malthus: Living within limits. *The Social Contract*, 181-187.

Hunter, L. M. (2001). *The environmental implications of population dynamics.* Santa Monica, CA: RAND.

Hunter, L. M., Twine, W., & Johnson, A. (2005). *Population dynamics and the environment: Examining the natural resource context of the African HIV/AIDS pandemic.* Institute of Behavioral Science, Research Program on Environment and Behavior, University of Colorado at Boulder, Boulder, CO, Working Paper.

Imathiu, I. (1998). Experts decry pollution of water source. *Daily Nation on the Web*. Retrieved November 21, 2006, from http://www.nationaudio.com/ News/DailyNation/1998/020798/Features/XX13.html.

Kirkland, T., Hunter, L. M., & Twine, W. (2005). *"The bush is no more": Insights on institutional change and natural resource availability in rural South Africa*. Institute of Behavioral Science, Research Program on Environment and Behavior, University of Colorado at Boulder, Boulder, CO. Working Paper.

Kunstler, J. H. (2005). *Ns essay. New Statesman, 18*(870), 23-25.

Landry, P. (2006). *Biographies: Thomas Robert Malthus* (1766-1834). Retrieved November 14, 2006, from http://www.blupete.com/Literature/ Biographies/Philosophy/Malthus.htm.

Malthus, T. (1803). *An essay on the principle of population as it affects the future improvement of society* (2nd ed.). London: J. Johnson.

Malthus, T. (2004/1798). *An essay on the principle of population, as it affects the future improvement of society, with remarks on the speculations of Mr. Godwin, M. Condorcet, and other writers* [Electronic version]. eBooks@Adelaide, retrieved November 17, 2006, from http:// etext.library.adelaide.edu.au/m/malthus/thomas/m26p/index.html. Original published 1798 in London by J. Johnson.

Maher, T. M. (1995). *Media framing and salience of the population issues: A multimethod approach*. Doctoral Dissertation, The University of Texas at Austin, Austin, Texas.

Meyerson, F. A. (2004). Population growth and deforestation: A critical and complex relationship. *Population Reference Bureau*. Retrieved November 20, 2006, from http://www.prb.org/Template.cfm?Section= PRB&template=/ContentManagement/ContentDisplay.cfm&Content.

Mugwang'a, M. (2006, November 9). Thousands flee their homes as slum death toll goes up. *Daily Nation Online*. Retrieved November 15, 2006, from http://www.nationmedia.com/dailynation/nmgcontententry.asp ?category.id=1&newsid=85169.

NATION Team. (2006, November 14). Violence blamed on politicians. *Daily Nation Online*. Retrieved November 15, 2006, from

O'Rourke, P. J. (1995). *All the trouble in the world: The lighter side of over-population, famine, ecological disaster, ethnic hatred, plague, and poverty*. New York: Grove/Atlantic, Inc.

The Race Relations Reporter. (2006, November 22). *Weekly Bulletin*. Forwarded by Bruce King of the USD Academic Affairs and Diversity, email, baking@usd.edu.

RAND. (2000). Population and environment: A complex relationship. *Population Matters: Policy Brief.* Retrieved November 20, 2006, from http://www.rand.org/pubs/research_briefs/RB5045/index1.html.

Smith, A. (1776). *An inquiry into the nature and causes of the wealth of nations* [Electronic version]. Retrieved November 17, 2006, from http://socserv2.socsci.mcmaster.ca/~econ/ugcm/3113/smith/wealth/wea/bk01&bk04. Original published in Oxford by Clarendon Press.

Tessler, M. (1994). *A history of the Israeli-Palestinian conflict.* Indianapolis, Indiana: Indiana University Press.

United Nations Fund for Population Activities. (2007). *State of the world population 2007: Unleashing the potential of urban growth* [Electronic version]. Retrieved June 27, 2007, from http://www.unfpa.org/swp/2007/english/introduction.html.

World Bank. (n.d.). *Poverty Maps* [Electronic version]. Retrieved November 20, 2006, from http://www.worldbank.org/research/povertymaps/kenya/ch5.3.3.pdf.

Chapter 3
Global Warming and Climate Change

Related to the problems of over population, environmental destruction, poverty, diseases, and other problems discussed in the previous two chapters is the issue of global warming and climate change due to human activities. As mentioned earlier, in 1974, James Lovelock and Lynn Margulis published a paper explaining their observation that the earth, together with its atmosphere, biosphere (the thin earth's crust that supports biological life), the oceans, rocks, etc. acted as a single system, regulating its own temperature and chemical composition.

Evidence of this self-regulation included the observation that the earth's temperature seemed to stay within a narrow range between freezing and boiling points. Other evidence was that the composition of atmospheric gasses seemed to be very stable. For example, the amount of oxygen was always about 21% in proportion to other atmospheric gasses. If it went below 15%, animal forms would not survive. If it increased to about 35% or more, there would be a global explosion. Ignited forest fires would be uncontrollable and would consume most of the earth's continental forests. Also, nitrogen existed in gas form rather than as a nitrate as may be logically expected. All the above circumstances, it seemed, existed in order to preserve the earth in a form that is habitable to life. They concluded that the earth and all its constituent systems (atmosphere, oceans, upper rock layers, biosphere, and air) functioned as if it were a live, self-regulating system. They named this system Gaia (after the mythical Greek earth Goddess). Later, Lovelock (1979) published a book elaborating the Gaia hypothesis. By definition, Gaia "begins where the crustal rocks meet the magma of the earth's hot interior, about 100 miles below the surface, and proceeds another 100 miles outwards through the ocean and air to the even hotter thermosphere at the edge of space" (Lovelock, 2006, p. 15).

In his most recent work, Lovelock (2006) suggested that Gaia was sick, and human beings, like cancer that devastated the body, were the cause of her illness. In his view, the human activity of farming "abrades the living tissue of its skin" and "pollution is poisonous to it as well as to us" (p. 2). In a way, if the planet were a living, evolving, self-regulating organism (which Lovelock calls Gaia), one could analogically think of humanity as constituting one of its specialized organs (lets call the organ 'humanity'). This organ could serve a Gaia enhancing role, for example by protecting endangered species that are crucial to preservation of the ecosystem and therefore the health of the planet and all its life. Individual human beings in this conceptualization could be seen as cells of the organ, co-existing with other cells (flora and fauna, air, rocks, soil, etc.).

However, like body cells that become cancerous, human beings are overgrowing and destroying other Gaia 'cells' including its 'skin' (vegetation cover) and its 'breathing' organ (atmosphere). Just as normal cells that have become cancerous kill the victim, human beings are threatening to kill Gaia. In other words, instead of the organ (humanity) serving a Gaia enhancing role necessary for her preservation, it has become cancerous and in fact a danger to the whole planetary body.

In Lovelock's view, the greatest danger is human activities that lead to emission of large amounts of gasses such as Carbon Dioxide which have greenhouse effects. Such gasses, which are invisible, allow the heat from the sun into the earth, but like glass panes in a greenhouse, trap the heat so that it does not escape into space, leading to increased temperature levels, which cause generalized global warming. To understand the full implication of this phenomenon, I use the analogy of a house constructed of clear glass panes. Imagine being locked in such a house, with no air conditioning, in the middle of a sunny day with clear blue skies during the summer season. It would be extremely hot inside and it would not take long for you to die of heat exhaustion. Through emission of greenhouse gasses as we go about our daily occupations through which we earn a living and enjoy life, we can metaphorically think of ourselves as constructing a glass roof all around our planet.

In essence, we are creating a situation in which we are locking ourselves in a planetary greenhouse. We are already beginning to simmer in the heat trapped by that clear gaseous roof that we have constructed. Lovelock compares this heating of the eco-system (Gaia) to an organism being sick and having a fever. In other words, in his view, human activities have made Gaia sick and feverish, and she is beginning to fight back the disease through destructive events such as floods, hurricanes, etc. just like the human body would activate its immune system in an attempt to fight cancerous cells. If we do not do something quickly, we are headed for a very difficulty time indeed. He compares our current course to being in an airplane flying over an Ocean while the engines are about to fail and we are not even aware.

When the theory of Gaia was first proposed, many scientists dismissed it out of hand. Even currently, Lovelock and his adherents are sometimes ridiculed as indicated by a recent assertion that the Gaia hypothesis is "a hopeless mix of pseudoscience, bad science, and mysticism" (Pigliucci, 2005, p. 21). According to Pigliucci, although the earth is evidently a self-regulating complex system as Lovelock asserts, this does not justify analogically comparing it to a living organism. Accepting such an analogy would encourage sloppy thinking that would be pernicious to science because it would not advance any useful testable hypotheses that could lead to increased understanding of the planet.

The criticism of Gaia hypothesis not withstanding, more and more scientists started writing about one of the issues raised by Lovelock and colleagues, namely, the dangers of climate change, due to human activities, to the ecological balance necessary for survival of life on earth (Lamb, 1972; Schneider, 1989; Schneider & Londer, 1984; Tickell, 1986). This led to introduction of issues of climate change on the world agenda as indicated by the formation of the Intergovernmental Panel on Climate Change (IPCC) in 1989 by the World Health Organization (WHO) and the United Nations Environment Program (UNEP). These efforts led to the Kyoto agreement on climate change in 1997, where a treaty was signed requiring industrialized nations to reduce emission of greenhouse gasses to 5.2% below the 1990 levels by the year 2012 (Daynes & Sussman, 2005). By August 2005, 153 countries had ratified the agreement.

It is notable that the United States, which is "the world's largest energy consumer and its biggest polluter accounting for 25% of the industrial world's emissions" (Anonymous, 2005, p.8), withdrew from the Kyoto accords under President Geroge W. Bush's administration although Vice-President Gore had signed the treaty during the Clinton administration (Revkin, 2006). This brings us to the debate about global warming, its effects, and what needs to be done. In the rest of this chapter, the debate will be explored through the following questions: Is there global warming, and if there is, do human activities (occupational performance according to this book) have anything to do with it? Is global warming a problem and should we do something about it? What could be the fate of the earth and all its life if nothing is done?

Global Warming:
Is It Real and Are Human Activities Responsible?

There is a strong consensus in the scientific community that climate change and global warming are realities of our time. What is generating most controversy is whether human activities are in some way responsible of causing these phenomena. Some people insist that while it is true that the earth is getting warmer, this state has nothing to do with human activities (Revkin, 2006). They argue that the earth has gone through hot and cold cycles throughout its history, and we are simply in one of the hot cycles. I listened to one such argument by an author on a C-SPAN book T.V. program (unfortunately I do not remember the name of the speaker) who argued that similar warming has been observed in Mars and there are no human beings there.

This view was probably derived from the data recently derived from the Mars Reconnaissance Orbiter [MRO] (Covault, 2006). Analysis of the Mars Polar Ice Caps by the Obiter indicates a planet that has a dynamic weather system, where "over just the last 100,000 years, there has been a really dynamic history of changing climate [occurring every few hundred years] and recorded in layers of ice" (p. 4 of 4). Notice that there is no indication from this data that Mars is warming. So, it is not clear how it supports this speaker's views. What he did not mention, however, is that the data also indicated that there once existed in Mars diverse watery habitat that could have supported life. There was evidence that there were hot springs, wet soils, standing bodies of water, and gullies indicating that water once flowed on the planet's surface. Yet, right now, the planet is a desolate, lifeless desert.

We do not know exactly what happened in that planet. A science fiction writer could make up a story about intelligent beings that once occupied Mars, and through their reckless activities, caused the drying up of the planet and extinction of all its life. Fiction aside, the real question is: what exactly happened in Mars to dry up its water reservoirs? If we discount human activity as a significant influence on climate change, do we mean that irrespective of what we do, we are fated to end up like Mars, and therefore we might as well do nothing? This may be so, but I do not think that attitude would be very responsible. It would be comparable to those who ignore Malthus' warnings and see the danger of overpopulation as mere alarmism (see discussion of the issue in chapter 2).

As mentioned earlier, although there are individuals who dismiss global warming as a serious problem, scientific consensus seems to be that it certainly is a problem that must be soberly addressed. In my literature search while preparing to write this chapter, not many published articles written by notable scientists were found making the argument that there was no global warming. While most of the scientific discourses that I reviewed agreed that the planet was warming up, the debate seemed to be whether this temperature spike was purely a natural phenomenon or human activities (occupational performance) contributed significantly to the problem.

In order to make an informed opinion on this debate, I did an extensive search of literature on both sides of the argument using scholarly databases such as CINAHL, Cochrane Database of Systemic Reviews, Eric, LEXIS NEXIS, Proquest, Sociological abstracts, Wiley Interscience, and Web Science. I used key terms such as "human activity and global warming", "arguments about human contribution to global warming", "debunking global warming theory", "disproving global warming theory", etc. In this search, few articles proposing that human activity does not contribute to global warming were found. For example, in the Proquest database alone, there were more than 100 hits in the search. However, less than 1% of them made the argument that human activity does not contribute significantly to the phenomenon of global warming. Chief among those who disputed human responsibility in the etiology of global warming were two scientists, Nir Shaviv, an astrophysicist, and Jan Veizer, a geochemist (Avery, n.d.; Broad, 2006; Cash, 2003; Science Letter Editors, 2003).

Veizer (2002) used proxies (carbon and oxygen isotopes) from seashell fossils found at the bottom of the oceans to measure their ratios as a means of indirectly determining the temperature of the ocean and how much carbon was available in the oceans and in the atmosphere in the distant past. Based on his calculations, he concluded that the amount of carbon dioxide in the atmosphere about 4 billion years ago was not much different from what we have today. He stated that in fact, during the last ice age which was about 350 to 545 million years ago, the concentration of carbon dioxide in the atmosphere was about 10 to 15 times the current levels, yet the earth was frozen. These observations led him to conclude that carbon dioxide may be an amplifier but not a driver of climate change. Furthermore, he observed a periodicity in climate changes every 135 million years or so, suggesting that there are some other major drivers of the change in play.

The view that there are some other factors driving global warming and climate change is not particular to Veizer. Some have suggested that there are other times during the human history, such as the medieval times (800 to 1300 ACE), when it was as warm, if not warmer than today (Behreandt, 2006). Shaviv (Cash, 2003) attributed these periodic changes to bombardment of the earth with cosmic rays that occurs approximately every 143 million years. He explained that our galaxy, the Milky Way, is shaped in such a way that it has arms of dense concentration of millions of stars, spiraling out from the center of the galaxy but moving at different rates. As the solar system orbits the galaxy, every 143 million years or so, it passes through one of these arms. During those times, bombardment of the earth by high energy cosmic rays made of particles from exploding stars within the galaxy peaks. The particles cause formation of low-level clouds which block the warming effect of the sun and lower the earth's temperature. When the solar system is out of these galaxy arms and solar activity is highest, high-level clouds form. High level clouds allow more rays of the sun to reach the earth which causes the warming of the planet.

The two scientists recently published an article in the Geographic Society of America suggesting that their research indicates that carbon dioxide concentration in the atmosphere has very little effect on global warming, and that human activity contributes very little to this phenomenon. Therefore, they argued, "a significant reduction of the release of greenhouse gases will not significantly lower the global temperature, since only about a third of the warming over the past century should be attributed to man" (cited in Science Letter Editors, 2003, p. 1 of 2). Veizer and Shaviv seem to join others who see predictions of disasters such as "increased incidence of death and serious illness", "increased risk of damage to a number of crops", "decreased water resource quantity and quality" (Behreandt, 2006, p. 12), and increased frequency and intensity of hurricanes, flooding of coastal regions, etc. as a result of global warming as unwarranted alarmism. Their argument seems to go something like this: 1) Global warming is a natural phenomenon; 2) Human activity contributes very little if anything to the phenomenon; 3) Therefore, stop making alarmist predictions and let us go on with business as usual.

The above argument seems to be motivated mainly by economic concerns. This is apparent in the suggestion that the proposal to curb greenhouse gasses as mandated by the Kyoto accords is an effort "to curb a phantom menace that may not exist, especially when such efforts would have disastrous economic consequences" (Behreandt, p. 15). This conclusion has however been disputed by the majority of other scientists according to published literature identified from my search. For example, a group of Earth and space scientists from the Potsdam Institute for Climate Impact Research challenged the claims by Veizer and Shaviv on the basis that "their conclusions are ill founded" (Anonymous, 2006, p. 1 of 2). They took issue with the fact that the two scientists' reconstruction of ancient climatic patterns that they used to dispute the theory of human contribution to global warming is based on only 50 meteorite samples, which introduces the possibility of overgeneralization from insufficient data. Furthermore, they claimed that in some instances, the data were adjusted, sometimes by up to 40 million years, in order to fit the conclusions.

The scientists who disputed the primacy of terrestrial contribution to global warming also pointed out that the models constructed by Veizer and Shaviv, which were based on geological times (consisting of millions of years), were not suitable for explaining short term climate variations. They did not explain the spike in the Earth's surface temperatures particularly over the past 100 years. Rather, "volcanic eruptions, changes in solar activity, and the concentration in greenhouse gases" (p. 2 of 2) were more suitable explanations of temperature variations on a small time scale. Besides, even Veizer (2002) acknowledged that human activity could not be completely discounted. While he argued that emission of greenhouse gasses was probably not the primary driver of climate change, particularly global warming, he conceded that it was certainly an amplifier of the phenomenon.

This view was supported by other scientists who argued that although the warming/cooling cycles of the planet were normal fluctuations, human activities were accelerating them (Behreandt, 2006). The latest report from the IPCC (2007) indicated that there was a high level of certainty (by about 90%) that human activities that led to accumulation of greenhouse gasses such as carbon dioxide, methane, and nitrous oxide into the atmosphere were primary drivers of climate change in the last 200 years (Collins, Colman, Haywood, et al., 2007). According to the report, since continuous environmental monitoring began in the 1850s, the highest concentration of CO_2 was recorded in the atmosphere in the last 10 years than at any other 10 year period (most of the accumulation precipitated by burning of fossil fuels). This dramatic accumulation of CO_2 in the atmosphere, they argued, could not be attributed to chance. Its concentration is currently at 35% above the pre-industrial levels. In the same period of time, methane gas concentration in the atmosphere (from agricultural activities and burning of fossil fuels) has increased by 2 1/2 times, and nitrous oxide concentration has increased by about 20%. Most of these increases in concentration have been in the industrialized heavily populated northern hemisphere.

In the same time period, temperatures and sea levels rose significantly. The last 11 out of 12 years have been the warmest since reliable record keeping of weather indicators began in 1850 (IPCC, 2007). It is difficult to see how 11 out of 12 consecutive years can be warmest purely by chance. Since 1993, the sea-level has risen at the rate of about 3.1 millimeters per year, mostly due to expansion of water as a result of this sustained planetary warming. Precipitation also increased in certain regions of the world while other regions became significantly drier and arid. All the above indicators suggest that human activities correspond significantly with increased concentration of greenhouse gasses in the atmosphere and subsequent climate changes (positive radiative forcing, or enhanced warming [IPCC, 2007]).

All the above evidence of negative anthropogenic influence on the climate not withstanding however, the question is: even if it were true that global warming is a natural phenomenon which will eventually happen anyway, should we ignore the human contribution which accelerates the process and stick to business as usual, or should we try to slow it down for the sake of not only humanity but all other life on the planet? My argument is that not doing anything would be highly irresponsible. It would be comparable to the argument that since we know that we are destined to die eventually, we should not seek remedy when we get ill. After all, death is a natural process. So, why not continue to engage in behaviors such as eating unhealthy food, engaging in dangerous sex, etc. even though such behaviors are likely to hasten our journey to death, which is after all our final destiny? It seems to me that just as a doctor would advise you to adopt a healthy lifestyle which is likely to prolong a quality life for you, similarly, any human activity that hastens the process towards destruction of life on the planet should be curtailed.

Significance of Global Warming and Climate Change

As discussed earlier, there are those who see the predicted consequences of global warming such as rise in incidences of human death and diseases, flooding of Coastal regions, destruction of crops, shortage of water and other resources as mere exaggerations and alarmist (Behreandt, 2006). We will now examine these predictions more closely and see whether there is any chance that they may have some merit.

Are There Increased Incidences of Death and Serious Diseases as a Result of Global Warming?

First, it is important to point out that the effect of global warming is not manifested uniformly around the world. Those in the Northern hemisphere like the US who may see no problem can be excused. After all, it will be a while before they really feel the full impact, although the deaths of several people in the summer of 2003 due to a heat wave in Europe suggest different[ii]. There is also evidence that climate change resulting from warming of the sea-surface temperatures (SSTs) due to man-made greenhouse gasses has led to increasingly more frequent, intense, and deadly cyclones (also known as hurricanes or typhoons) as recorded by scientists in 2004 and 2005 (Trenberth, 2007). Furthermore, there have been frequent reports of severe flooding in parts of the USA and Europe[iii]. Many people experiencing these events state that they have never seen anything like it in their lives. However, While the average change in global temperature over the past century has been about 1 degree Farenheit (0.6 degrees Celsius) (Coren, 2006; Trenberth, 2007), in places like Kenya, there has been an increase by up to 3.5 degrees Celcius in some regions in the last 20 years (Mwango, 2006). This is by all means a significant increase.

The concrete effect of the change in temperature has been an increase in cases of malaria. Mosquitoes used to do well in the lower altitudes. However, within the last 20 years, more and more people in the Kenyan Highlands such as Nyeri and Karatina have been diagnosed with malaria. This is because the temperatures in those areas are now conducive to the survival and breeding of the female (anopheles) mosquito which carries the malaria parasite. The anopheles mosquito does well in warm areas (temperatures between 20 and

30 degrees Celcius). Since the country's Highlands warmed up by between one and three degrees Celcius since the 1990s, the average temperature ended up being about 18 to 21 degrees Celcius, which made the region conducive to mosquitoes.

Consequently, a record 20 million Kenyans a year (54,795 daily) are at risk of contracting malaria. Available data indicates that 12,329 cases of malaria are being reported every day, an increase of incidence by 50 to 100%. These statistics are consistent with the findings from a survey conducted by the World Health Organization and the London School of Hygiene and Tropical Medicine in which it was found that "the 'ancillary effects of climate change, such as *malaria* and malnutrition, are causing approximately 160,000 deaths each year - *more than terrorism*'" (Daynes & Sussman, 2005, p. 438, emphasis mine). These data do not include other deaths caused by diseases such as cholera and other diarrheal conditions which are related to severe flooding as was occurring in the country of Kenya during the December 2006/January 2007 rain season, which could also be attributed to global warming as will be seen later. In addition, there is evidence that respiratory and skin diseases associated with increasing temperatures are on the rise in the country (Mwango, 2006). Therefore, there is convincing evidence supporting the prediction that global warming is associated with increased incidences of death and diseases.

Is There Flooding of Coastal Regions?

In the year 2005, we witnessed a sad phenomenon of hurricanes Katrina and Rita in the USA which caused havoc in the southern coastal region. The city of New Orleans was virtually decimated and many people lost their lives. Experts were quick to come on television and discount any connection between these events and global warming (see Limbaugh, 2005). However, we should remember that one of the predictions of the so called alarmist scientists (mainly those associated with the UN's Intergovernmental Panel on Climate Change [IPCC]) was that with increased temperature on the earth's surface, hurricanes and floods would increase both in frequency and intensity (Barasa, 2006; Behreandt, 2006). Therefore, the destructive intensity of Katrina and Rita had been predicted. In 2007, it has been predicted that there will be a record number of major storms in the Americas. We already saw a category 5 hurricane (hurricane Dean) which battered the

Island of Jamaica and the Yukatan Peninsula in Mexico last August, 2007. More such storms are expected before the season is over.

We also see increased frequency and intensity of flooding, landslide damage, mudslides, etc. in many places as predicted by the IPCC. In Kenya for instance, there has been two incidences of excessively heavy rains with consequent disastrous flooding all over the country twice in less than 10 years (around 1998, and 2006). The heavy rains that occurred in December 2006/ January 2007, and which caused much damage and suffering all over the country, were highly unusual. There are usually no rains in Kenya in late December and January.

This aberration in weather was happening in other parts of the world as well. On December 3rd, 2006 for example, the province of Albay in the Philippines was hit by an intense typhoon which caused mudslides that uprooted trees, flattened houses, and drowned people (Marquez, 2006). It was estimated that as many as 1000 people lost their lives in that catastrophe. Gonzalez, the Governor of the Province stated: "Never before in the history have we seen water like this. Almost every residential area was flooded" (p. 2 of 2). Some may see these events as proof that the IPCC scientists were correct in their predictions. One may argue that indeed, incidences of hurricanes and other types of dangerous weather events are becoming more frequent and intense as the global temperature rises.

Is There Destruction of Crops?

The consequences of global warming include not only destruction of crops but also all kinds of infrastructure (Barasa, 2006). It was ironic that the floods that were destroying roads and other types of infrastructure in addition to crops in Kenya in 2006/2007 were happening while the UN Framework Convention on Climate Change conference was taking place in Nairobi. We are also familiar with the destruction of infrastructure that happened in New Orleans as a result of hurricane Katrina in 2005. Besides infrastructure and food crops, there is evidence that other forms of life are being affected. For example, it has recently been documented that the migratory behavior of many bird species is now affected by global warming and these species, all over the world, are being threatened with extinction (Nation Media Group, 2006). As an aside, it is instructive that birds (specifically Canaries) are used in places such as the coal mines to detect concentrations of Carbon Monoxide before humans are exposed. This is because they are highly sensitive to changing environmental conditions. These same birds are now being affected, indicating that there are already some dangerous climatic shifts. How long before human beings begin succumbing to the changes?

Is There a Shortage of Water as a Result of Global Warming?

Again, there is no discernible water shortage yet in the world hemispheres. However, we should not forget that there are dangerous trends taking place. In 2002, satellite imaging showed that there was increasing instability in the Larsen B ice shelf in the Antarctic due to melting of polar ice (Behreandt, 2006; See also the award winning documentary by Gore, 2006). Similarly, in Greenland, about 57 cubic miles of ice are being lost every day from the massive ice-sheet in the region. The serious real life effects of these changes are most pronounced in the Equatorial region. For example, Lakes Nakuru, Naivasha, Elmenteita, and Victoria in E. Africa are drying up (Barasa, 2006). The ice caps on Mt. Kilimanjaro (the highest mountain in Africa) and Mt. Kenya are melting.

These are important sources of water for the East African region. Therefore, these events are extremely worrying. We may be headed for severe human crisis in the region in the near future if the trend continues and these sources of water completely dry up. As Sachs (2007) warned, we may be about to experience the problem of climate change-related refugee problem where millions of people may be compelled to relocate. Already, places like Turkana have been hit by severe droughts in recent times, for three consecutive years to be exact, prompting Mr. Miliband, the UK's Secretary for Environment who visited the region to state that "climate change is already here and a reality" (Barasa, 2006, p. 1 of 2). We also know, as stated in the previous chapter, that the conflict in Darfur, Sudan, is in part related to increasing water shortages (see discussion in chapter 2). Sachs (2007) suggested that: "The violence in Darfur and Somalia is fundamentally related to food and water insecurity. Ivory Coast's civil war stems, at least in part, from ethnic clashes after people fled the northern drylands of Burkina Faso for the coast" (p. 43). This is where the interaction between population growth and climatic/environmental changes becomes most apparent.

The above discussion leads to the conclusion that contrary to those who argue that global warming and climate change do not pose a real problem, the predictions of human suffering due to increased incidences of death and disease, flooding, mudslides, and water shortage are phantom problems, and that we need do nothing to curb the problem, are wrong. There is much evidence that vindicates the predictions of IPCC. The effects of global warming are already affecting many people, particularly around the tropics. In those regions, diseases related to increased planetary temperatures are on the rise, the frequency and intensity of both flooding and droughts are significantly increasing, and water sources in the region are drying up. The problems we are dealing with are best summarized by Savage (2006) thus: "The polar ice caps are melting, summers are growing hotter and hurricanes are becoming more powerful..." (p. 5A).

What Should We Do About the Problem?

If indeed global warming is a real problem as indicated in the discussion above, what should we do about it? It seems that again there are two points of view regarding this issue. There are those who think it should be business as usual, because global warming is a "phantom menace" and we do not need to do anything about it (Behreandt, 2006, p. 15). Such views are supported by scientists such as Veizer and Shaviv who argue that human activities contribute very little to the problem because most of it is a natural process related to terrestrial activities, and therefore cutting back on emission of greenhouse gasses, which has been proposed as a solution, will not curb the trend (Anonymous, 2006; Science Letter Editors, 2003).

Others, and these seem to be the majority as indicated by literature found in my search (see earlier discussion of literature search for this chapter), support the IPCC (2007) view that global warming constitutes an emergency that needs to be addressed as a matter of priority (Anonymous, 2005; Barasa, 2006; Daynes & Sussman, 2005). The most popularly suggested approach to the solution of the problem according to those who feel that something should be done is to take measures to curb industrial emission of greenhouse gasses into the atmosphere. Perhaps the best effort in this endeavor is the Kyoto accords which were discussed earlier. Based on analysis of air bubbles trapped in ancient ice, these scientists argue that the levels of greenhouse gasses in the atmosphere today are higher than at any time in at least the last

420,000 years (Coren, 2006). This denotes that we have to reduce such emissions drastically and immediately. Therefore, as discussed earlier, the Kyoto agreement calls for industrialized nations to cut down emission of such gasses to at least 5.2% below the 1990 levels by the year 2012 (Anonymous, 2005; Daynes & Sussman, 2005; Revkin, 2006). Al Gore signed the accords but when Bush assumed office in 2001, he withdrew from the treaty. The Bush administration has strongly resisted suggestions to curb greenhouse gasses even though the United States alone is responsible for at least 25% of those gas emissions.

The third approach is suggested by those in Bush's camp, who it seems do not believe in the danger of global warming but want to make a token effort in response to the pressure to do something. I say they do not seem to believe because their attitude towards the whole problem seems to be captured by Behreandt (2005) in the following statement:

> Constrained since taking office by *popular conservative disdain of global warming alarmism*, the Bush administration has had to steer the nation away from the UN's Kyoto accord. That treaty mechanism, *favored by leftists in Europe and elsewhere*, would have required reductions in emissions of carbon dioxide and other *so-called greenhouse* gases to such a degree that the U.S. economy would have been severely harmed. By one widely cited estimate, the cost of implementing Kyoto would be a staggering $716 billion. (p. 19, emphasis mine)

Terms such as the *so-called greenhouse* gases demonstrate the dismissive attitude by those who favor doing nothing about the problem. Even the cost of implementing the Kyoto accords cited above is a mere token compared to the amount of money being spent in the Iraq war every week (currently amounting to trillions of dollars since the war began). Furthermore, statements such as the above make one wonder whether opponents of measures to reduce global warming are mostly pro-big business advocates whose only concern is to make a profit at any cost. Obviously, one can see that they have much financial motivation to resist any attempt to curb greenhouse gases because that would mean corporations taking serious steps to change the way they do business and trying to conduct their affairs in an environment-friendly manner.

Never the less, perhaps due to pressure from many scientists who really see a problem, the Bush administration came up with its own proposal to reduce environmental pollution in general (King, 2001; Wallace, 2002). In his version, Bush proposed that reduction in greenhouse gasses should be pegged on economic output, such that there would be a reduction of the ratio of greenhouse gas emission to economic output of 18% over 10 years. This translated into about 4.5% actual reduction in 10 years. It seemed that this was just a token measure meant to pacify the World by making everybody believe that they were actually doing something about the problem.

Therefore, as can be seen in the discussion above, there seems to be four views regarding how we should respond to the problem of global warming: 1) do nothing because global warming is really a non-problem; 2) do nothing, even though evidently there is global warming, because reduction of greenhouse gases will not curb the process since its driven by terrestrial factors rather than human activity; 3) reduce green-house gas emissions significantly as a matter of priority because such gases are a significant amplifier of the process of global warming if not outright causes; and 4) show an effort to reduce the greenhouse gases for Public Relations (PR) purposes to make peace with those who want something done even though there is really no reason to do anything.

Of course there is also a fifth, less mainstream opinion. Lovelock (2006) believed that we had already destroyed the environment so much that sustainable development was no longer an option. In his view, reducing emission of greenhouse gases was not going to solve the problem. According to him, what was called for was cutting out such emissions all together. However, because our civilization was dependent on sources of energy that inevitably led to accumulation of such gases in the atmosphere, what we needed, he argued, was strategic retreat rather than sustainable development. In this retreat, he argued, nuclear energy was crucial. Such a source of energy was not as dangerous as it had been made out to be, and if we started using it, we would be able to keep the economies that drive contemporary human civilizations going as we explored other environment friendly energy sources.

In addition, Lovelock argued, we could not afford to destroy any more of the natural land by converting it to farmlands because too much of the earth's natural environment had already been destroyed and it needed time to heal. He suggested that in addition to adopting the use of nuclear energy, we should find ways of synthesizing food from chemical elements in the atmosphere so that we did not need to use more land to grow food to feed the growing human population (in addition, I would recommend that aggressive programs to curb human population growth be initiated in order to limit the pressure exerted on the environment by the number of people on the planet).

Furthermore, he continued to argue, we could not wait for consensus in order to act to save our planet. We needed to begin acting locally, so that each nation individually committed to taking the initiative rather than waiting for consensus between the world powers such as the USA before we could do anything. As will be discussed in part 2 of this book, I would argue that we need to go even more local than that. It will be argued that the reason why any proposed measures to curb environmental destruction do not succeed too well is because we approach the issue largely in a deductive manner. We assume that treaties such as the Kyoto accords will translate into modification of individual corporations and even individual personal behaviors.

Instead, I will argue that we need both deductive and inductive approaches to the problem. Deductively, governments need to make a real commitment to facilitate changes that will save life on the planet. Inductively, individuals need to be sensitized so that they realize that their individual choices as they go about their daily occupations contribute to environmental problems such as global warming and climate change. This conscientization would lead to people making more responsible choices and acting accordingly at the individual level. Furthermore, as mentioned earlier, we need to understand that addressing global warming and climate change cannot be effectively done without paying attention to the problem of over population. After all as an example: "According to one estimate, population growth will account for 35 percent of the global increase in CO2 emissions between 1985 and 2100 and 48 percent of the increase in developing nations during that period" (RAND, 2000, p. 4 of 6). Other issues that must be addressed contemporaneously include poverty, material inequalities, institutional functioning, etc. as discussed in the previous chapters.

What Might Be the Fate of Life on Earth if Nothing Is Done?

Scientists working with the UN's IPCC agree that by the year 2100, if the current trend does not change, global temperatures will rise by between 1.4 and 5.8 degrees Celcius ([Coren, 2006] note that in the 2007 IPCC report, these estimates were revised upwards based on the most recent data and computer models). This projection is already being realized. We have seen that in some places in Kenya, temperatures have risen by up to 3.8 degrees Celcius in less than 10 years. However, many scientists also agree that it would be very difficulty to predict what will happen beyond the year 2100. This inability to forecast what will happen far into the future is a consequence of the fact that the earth (or to borrow from Lovelock, Gaia) is a complex system.

There are many variables involved, which are in dynamic interaction with each other. For example, weather patterns are constantly changing and interacting with human made changes, as well as with other life forms that are also constantly changing in their endeavor to compete for natural resources and ultimate survival. These variables make a planetary system that is moving through space and interacting dynamically with the external environment. Moreover, the interaction between variables is non-linear, making predictive modeling even more difficulty. Given this complexity, it is almost impossible to predict how the system will behave far into the future.

However, we also know that complex systems are sensitive to initial conditions (Crutchfield, Farmer, Parkard, et al., 1995; Mouck, 1998; Waldrop, 1992), or what may be called local conditions. This means that small changes in local interaction between variables may translate into big differences in system trajectory outcomes. That is why, if we agree that the earth is a complex planetary system, we cannot discount the effect of generation of greenhouse gases due to human activity as Veizer and Shaviv suggest. As an illustration, it has been suggested that a volcanic explosion on Mt. Tambora on the small highland of Sumbawa in Indonesia in 1815 resulted in massive amounts of rock and dust being propelled 150 cubic kilometers into the higher atmosphere (Tkachuck, 1983). This event caused massive cooling in Europe and Eastern USA in 1816 (referred to as "the year without summer"), resulting in death of thousands of people and animals due to diseases and starvation, and loss of thousands of livelihoods.

Apparently, the mechanism that caused this little ice age was simple. Some of the constituents of volcanic emissions are sulfur particulates. These particles (aerosols) absorb the sun's rays causing a general cooling of the earth. Furthermore, sulfur combines with moisture forming droplets of sulfuric acid causing clouds that hang around in the upper atmosphere for years. These clouds block the sun's rays causing further cooling of the earth. Other aerosols from the eruption that were accumulated in the atmosphere reflected the sun's rays into space causing more cooling [a phenomenon known as negative radiative forcing] (Collins et al., 2007).

In other words, a local event in a small island in Indonesia caused climate changes that were global and disastrous to thousands of people, who were probably not even aware of the event. If a single local event can result in such far reaching global effects, what makes us think that continually pumping pollutants into the atmosphere is not going to have any consequences that affect us? We do not know what factor may push the system over some unseen threshold and cause systemic dis-equilibrium leading to self-reorganization at a different level of functioning. It could be the next car, or the next industrial complex, or the next human being who is born, who tips the system over.

If the threshold is crossed, the planetary system would go into self-organization. This self-organization may involve increased temperature ranges where the difference between the cold and hot weather is extreme (for instance temperatures varying between negative 200 and positive 200 degrees Celcius in the same day), the composition of gases in the atmosphere may be significantly different such that human beings and other animals are unable to breath, etc. In these extreme circumstances, life as we know it today would probably not be viable. This need not be bad for the planet. Human beings and other life forms need the planet, but the planet does not necessarily need them. Through self-organization, it may transition into another level of self-regulation, and human beings and other life forms may not be required for this new level of self-regulation.

This process could begin, as the IPCC scientists predict, with increased incidence of death and diseases, flooding of Coastal regions, more frequent and intense hurricanes, typhoons, more frequent draughts and other natural phenomena, destruction of crops, drying up of water sources, etc. As the process accelerates, most of the living things would go into extinction and eventually, earth would be a desolate, life-less, desert, like her sisters Mars and other planets in the solar system. There are even those who suggest that global warming may paradoxically lead to cooling due to disruption of the Atlantic Thermohaline Circulation as the polar ice melts (Behreandt, 2006; Gore, 2006) precipitating another ice age with equally calamitous effects on human civilization. But then again, may be the opponents of the global warming theory are right and nothing will happen. The question is, do we want to take a chance and find out the hard way or would it be more prudent to do our best to slow down the process, even though we may not be sure of what the exact process may be?

Reflection Exercise #3

After reading chapter 3, answer the following questions:

1. Before reading chapter 3, how aware were you (on a scale of one to ten with one being not at all aware and ten being completely aware) about the fact that:

    a. The earth is getting warmer?
    b. There is a debate about whether or not human activities are to blame for the increasing earth's temperature?
    c. The consensus by the majority of scientists in the world is that human activities that lead to accumulation of greenhouse gasses in the atmosphere are to blame for increasing earth's temperature?
    d. Diseases such as malaria are becoming more prevalent and fatal as a result of increasing global temperatures?
    e. The arctic ice is melting and so are the ice caps in Mount Kenya and Mount Kilimanjaro?
    f. Lakes such as Victoria in East Africa are drying up?
    g. Natural calamities such as tornadoes, floods, draughts, etc are becoming more frequent and intense?

2. What do you think could be the fate of life on earth if the problem of global warming is not curbed?

3. What can you do to become more informed about the problem of environmental destruction and global warming/climate change, and associated problems?

4. What can you do as an individual to contribute towards curbing the problem of environmental destruction and global warming/climate change?

References

Anonymous. (2006). Role of cosmic rays in climate change refuted [Electronic version]. *Bulletin of the American Meteorological Society, 85*(4), 490-491. Retrieved November 30, 2006, from http://proquest.umi.com/pqdweb?index=2&sid=2&srchmode=1&vinst=PROD&fmt=3&s...

Anonymous. (2005). A stepping stone. *Canada and the World Backgrounder, 71*(2), 8-13.

Avery, D. (n.d.). Seashells say earth temperatures driven by cosmic rays, not CO_2. *Center for Global Food Issues*. Retrieved November 30, 2006 from http://www.cgfi.org/materials/articles/2003/aug_22_03.htm.

Barasa, L. (2006, November 14). Climate change to blame for current devastating rains [Electronic version]. *Daily Nation Online*: Special Report. Retrieved November 14, 2006, from http://www.nationmedia.com/dailynation/nmgcontententry.asp?category_id=1&newsid=85648.

Behreandt, D. (2006). Global warming: Too hot or not? *The New American, 22*(19), 10-15.

Behreandt, D. (2005). Building the post-Kyoto future. *The New American, 21*(23), 19-21.

Broad, W. J. (2006, November 7). In ancient fossils, seeds of a new debate on warming. *New York Times*, F.1.

Cash, S. (2003, December 1). Cosmic heating [Electronic version]. *The Jerusalem Report*. Retrieved November 30, 2006, from http://proquest.umi.com/pqdweb?index=8&sid=2&srchmode=1&vinst=PROD&fmt=3&s...

Collins, W., Colman, R., Haywood, J., Manning, M. R., & Mote, P. (2007, August). The physical science behind climate change. *Scientific American*, 64-73.

Coren, M. (2006, February 10). The science debate behind climate change [Electronic version]. *CNN*. Retrieved November 20, 2006, from http://www.cnn.com/2005/TECH/Science/04/08/earth.Science/index.html.

Covault, C. (2006, October 23). Aviation week & space technology. New York [Electronic version]. *World News & Analysis*. Retrieved December 1, 2006, from http://proquest.umi.com/pqdweb?index=3&sid=1&srchmode=1&vinst=PROD&fmt=3&st...

Crutchfield, J., Farmer, J., Parkard, N., & Shaw, R. (1995). There is order in chaos: Randomness has an underlying geometric form. Chaos imposes fundamental limits on prediction, but it also suggests causal relationships where none were previously suspected. In R. Russell, N. Murphy, & A. Peacocke (Eds.), *Chaos and complexity: Scientific perspectives on divine action* (pp. 35-48). Berkeley, CA: The Center for Theology and the Natural Sciences.

Daynes, B. W., & Sussman, G. (2005). The "Greenless" response to global warming. *Current History, 104*(686), 438-443.

Gore, A. (Writer and Actor), West, B. (Actor), & Guggenheim, D. (Director). (2006). *An inconvenient truth* [Motion picture]. (Available from Paramount Classics, 2025 Broadway, Oakland)

Intergovernmental Panel on Climate Change. (2007). *Climate change 2007: The physical science basis: Summary for policymakers.* Geneva, Switzerland: IPCC Secretariat.

King, J. (2001, June 11). Bush offers alternative environmental plan [Electronic version]. *CNN: Inside Politics.* Retrieved November 28, 2006, from http://archives.cnn.com/2001/ALLPOLITICS/06/11/bush.global. warming/index.html.

Lamb, H. (1972). *Climate: Present, past, and future.* London: Methuen.

Lovelock, J. (2006). *The revenge of Gaia: Earth's climate crisis & the fate of humanity.* New York: Basic Books.

Lovelock, J. (1979). *Gaia: A new look at life on earth.* Oxford: Oxford University Press.

Lovelock, J. E., & Margulis, L. (1974). Atmospheric homeostasis by and for the biosphere: The gaia hypothesis. *Tellus, 26,* 2-9.

Limbaugh, R. (2005, September 27). It's solar warming, not global warming. *The Rush Limbaugh Show.* Retrieved November 29, 2006, from http://www.rushlimbaugh.com/home/eibessential/estack/enviro-wacko-Update.guest.html.

Marquez, B. (2006, December 3). Red Cross: Asia storm toll may hit 1,000 [Electronic version]. *USA Today.* Retrieved December 3, 2006, from http://www.usatoday.com/weather/news/2006-12-02-Phillippines-typhoon_x.htm?csp=24.

Mouck, T. (1998). Capital markets research and real world complexity: The emerging challenge of chaos theory. *Accounting Organizations and Society, 23*(2), 189-215.

Mwango, D. (2006, November 8). Global warming blamed for malaria upsurge [Electronic version]. *Daily Nation Online.* Retrieved November 8, 2006, from http://www.nationmedia.com/specials/climate/nairobi/nairobi0611064.htm.

Nation Media Group. (2006, November 14). Global warming could wipe out most birds - WWF [Electronic version]. *Daily Nation Online*. Retrieved November 15, 2006, from http://www.nationmedia.com/specials/climate/nairobi/nairobi1411064.htm.

Pigliucci, M. (2005). *The so-called Gaia hypothesis*. *The Skeptical Inquirer, 29*(3), 21.

RAND. (2000). Population and environment. *Population Matters: Policy Brief*. Retrieved November 20, 2006, from http://www.rand.org/pubs/research-briefs/RB5045/index1.html.

Revkin, A. C. (2006). Is our planet in peril? *New York Times Upfront, 139*(1), 20-23.

Rucker, B. (2007, August 17). Heat wave kills 41 in South, Midwest [Electronic version]. *Associated Press*. Retrieved August 17, 2007, from http://www.topics.net/content/ap/2007/08/heat-wave-kills-41-in-South-midwest?fourSS=1.

Rush On Line. (n.d.). Global warming. Retrieved November 29, 2006, from http://www.rushonline.com/visitors/globalwarming.htm.

Sachs, J. D. (2007, June). Climate change refugees: As global warming tightens the availability of water, prepare for a torrent of forced migrations. *Scientific American*, 43.

Savage, D. G. (2006, November 26). High court to rule on pollution policy: Federal government should regulate vehicle and plant emissions, states say. *Argus Leader*, 1A.

Schneider, S. H. (1989). *Global warming*. San Francisco: Sierra Club Books.

Schneider, S. H., & Londer, R. (1984). *The coevolution of climate and life*. San Francisco: Sierra Club Books.

Science Letter Editors. (2003, September 1). Climatology; global warming not man-made phenomenon, researchers say [Electronic version]. *Science Letter*, Atlanta. Retrieved November 30, 2006, from http://proquest.umi.com/pqdweb?index=9&sid=2&srchmode=1&vinst=PROD&fmt=3&s...

Tickell, C. (1986). *Climate change and world affairs*. Cambridge, Mass.: Harvard University Press.

Tkachuck, R. D. (1983). The little ice age [Electronic version]. *Origins, 10*(2), 51-65. Retrieved December 6, 2006, from http://www.grisda.org/origins/10051.htm.

Trenberth, K. E. (2007, July). Warmer oceans, stronger hurricanes: Evidence is mounting that global warming enhances a cyclone's damaging winds and flooding rains. *Scientific American*, 45-49.

Veizer, J. (2002). Global warming: Armageddon or bust? *Forum for Applied Research and Public Policy, 16*(4), 57-63.

Waldrop, M. M. (1992). *Complexity: The emerging science at the edge of order and chaos*. New York: Simon & Schuster.

Wallace, K. (2002, February 14). Bush to unveil alternative global warming plan [Electronic version]. *CNN*. Retrieved November 28, 2006, from http://archives.cnn.com/2002/ALLPOLITICS/02/13/bush.global.warming/index.html.

PART II
INDIVIDUAL OCCUPATIONAL PERFORMANCE: PART SOLUTION TO GAIA'S PROBLEMS

In Part I (chapters 1 to 3) some of the prevalent global problems of our time were highlighted. They include extreme poverty, diseases, material inequalities, dysfunctional social/political/cultural institutions, environmental destruction, and global warming/climate change. It is important to make explicit at this point an important assumption in this book, which is hopefully already apparent, that all the above listed issues are inter-related and none of them can be resolved in isolation. This view is consistent with many modern scientists' perspective, such as Capra (1996) who held the position that in the modern holistic view, the world was an integrated whole and not a collection of discreet parts.

Capra asserted that the major problems of our time "cannot be understood in isolation. They are systemic problems, which means that they are interconnected and interdependent" (p. 3, see also Capra, 1999). To illustrate briefly, as has become clear, overpopulation leads to environmental destruction which is related to global warming/climate change. It is also correlated with poverty as indicated by the fact that families with large numbers of children tend to be economically poorer compared with those with less number of children. Similarly, poorer countries tend to be characterized by high population growth and families with many children.

In addition, poor people may be less equipped to take care of the environment. It has already been pointed out that people with limited economic resources in the third world tend to depend on firewood for cooking and heating their homes because they cannot afford gas or electricity, and therefore are more likely to destroy trees contributing to deforestation. They are also more likely to be adversely affected by diseases. Corrupt dysfunctional governments are not effective in protecting the environment, meaningfully alleviating poverty, or decreasing material disparities within society. One can demonstrate the inter-connectedness among all the listed issues. Therefore, in order to solve any of them, all the others need to be addressed.

In section II (consisting of chapters 4 and 5), it will be argued that each of the problems identified in part I can be attributed to individual choices in daily occupational performance pursuits. Consequently, the issues can be resolved by sensitizing individuals about the consequences of their occupational performance choices. It is proposed that constructs from occupational science can be used to guide us in illuminating the consequential impact of individual occupational performance choices on the global issues discussed in the first three chapters.

In chapter 4, it will be proposed that occupational science may significantly contribute to the development of a conceptual framework to help develop programs to resolve the global issues in question. The scientific discipline of occupational science will be introduced and the ways in which it may contribute to the solution of pertinent issues will be explored. In particular, the constructs of occupation and occupational performance as they are understood in occupational science and occupational therapy will be defined as part of an attempt to demonstrate how occupational science can contribute to the solution of the identified global issues.

In chapter 5, it will be argued that each of the issues discussed in chapters 1 through 3 can be attributed to individuals' actions as they make choices and pursue daily occupations. This will be the basis of the argument in subsequent chapters that the solution of some of the problems facing us may be by sensitizing and educating individuals so that they can alter their occupational performance choices in order for their occupational behavior to contribute towards making the status of Gaia better rather than worse.

Reference

Capra, F. (1999). *The tao of physics: An exploration of the parallels between modern physics and Eastern mysticism.* Boston: Shambhala.

Capra, F. (1996). *A new scientific understanding of living systems: The web of life.* New York: Anchor Books/Doubleday.

Chapter 4
Occupational Science: A Source of Conceptual Framework for Partial Explanation of the Etiology of Gaia's Illness and Prescription for Healing

In the last 3 chapters, some of the global problems confronting us were outlined. These problems were postulated to be the symptoms of an ailing planetary ecological system, or what Lovelock (2006) called Gaia. In this chapter, the discipline of occupational science will be introduced as a possible source of a conceptual framework that can contribute significantly to the explanation (at least in part) of those symptoms by arguing for human occupational performance as a part of their cause. The conceptual framework will also be used to contribute to available prescriptions for actions that may help facilitate Gaia's healing. Using occupational science as suggested above is one way of launching it as a legitimate scientific discipline that can contribute significantly to the solution of wider societal issues. It is also consistent with the call for an attempt: "To reach a holistic understanding of human-altered ecosystems" (and other human systems that impact on the welfare of all people and all things on earth) through collaborative contribution by "ecologists and social scientists" in the pursuit for solutions (Heemskerk, Wilson, & Pavao-Zuckerman, 2003, p. 8).

Up to now, knowledge from occupational science has been used mostly in occupational therapy. Expansion of the scope of the discipline in the manner suggested above is consistent with the call for the discipline to contribute towards guiding the use of occupation *"for personal or social transformation"* (Townsend, 1997, p.18, emphasis mine), to facilitate engagement "in educational activities to promote a wider understanding of the occupational nature of humans..." (Wicks, 2000, p. 87), and to help individuals and communities realize their occupational potential (Wicks, 2005). The wider understanding referred to above may be taken as illumination of the role of occupation to the solution of global human concerns. Occupational potential has been defined as, "future capability, to engage in occupation towards needs, goals, and dreams for health, material requirement, happiness and well-being" (Wilcock, cited in Wicks, 2005, p. 130). To the above definition of occupational potential may be added the capability to enhance ecological well-being through occupation, as part of what Wilcock (2007) referred to as tapping into the potential of occupation to contribute to the improvement of the environment and the human condition.

Indeed, involvement of occupational science in resolution of the issues discussed in this book has been advocated previously by Wilcock (2006). She argued that humans were occupational beings who were intimately interconnected with other humans and all living things in an ecological context [what Naess, cited in Capra (1996) referred to as "deep ecology"]. In what she termed an *occupation-focused eco-sustainable development approach to health*, she suggested that occupational science should be involved in the process of empowering community members so that they were effectively involved in actions and policy decisions to facilitate a sustainable healthy relationship between people and their habits and lifestyles, communities, other living things, and environments. She saw this healthy relationship as achievable through a process of consultation and deliberation within communities in order to develop sound policies for resource management through ecologically health-enhancing occupations. In the present book, Wilcock's suggestions were developed further by providing specific guidelines regarding how individuals could consciously elect to change their occupational performance patterns in order to facilitate realization of healthy relationships between humans and their ecosystems.

Since the issues discussed in this book are interdisciplinary, it may be helpful to provide a brief introduction of occupational science so that those who may not know what it is may understand the arguments being advanced. Therefore, for the rest of this chapter, the discipline of occupational science and its proposed role in providing a conceptual framework for an occupation-based approach to understanding the global issues of our time will be discussed as follows: a brief historical development of occupational science will be provided; the discipline will be defined (including a definition of occupation as it is understood by occupational scientists) and its scope briefly described; an argument for the discipline as a source of conceptual framework to provide a new understanding of global issues discussed in chapters 1 to 3 and partial solution of the issues will be advanced.

A Brief Historical Overview of Occupational Science

Occupational science originated from occupational therapy and up to now, many of the proponents of the discipline are occupational therapists (Wilcock, 2001). As such, even though it is an independent discipline, it cannot be understood without reference to occupational therapy. The evolution of occupational science is part and parcel of the development of the idea of occupation as an integral factor in what might be understood as good health. Although according to Wilcock this idea can be traced back to ancient medical practices, particularly to the concept of *Regimen Sanitatis* which referred to a prescription of rules for living in order to help prevent illness and promote health, this historical overview will begin in the 20th century.

In 1917, six individuals from varying professional backgrounds and experiences met at Clifton Springs, New York, to found a new medical profession (Ikiugu, 2007; Quiroga, 1995). They were united by a strong belief in the healing power of daily occupations and the profession that they founded was meant to promote their use as a healing tool. One of the most influential leaders in the newly formed profession was Dr. William Dunton Jr. He formulated what could be understood as the first principles of the young profession. His view of the role of occupation in human well-being was apparent in his writings. For example he wrote that occupation was as important to life as food and drink (Dunton, 1919), implying that it was for all intents and purposes a basic need comparable to biological needs.

Through the guidance of these leaders, in the early years of the 20th century, the importance of both physical and mental occupations for holistic integration of mind and body in action leading to well-being was well recognized especially in rehabilitation medicine. Over time however, occupational therapy became more and more mechanistic and reductionistic (Ikiugu, 2007; Kielhofner, 2004). In the 1960s, 1970s, and 1980s, there was a call for the profession to return to the roots and rediscover the original principles emphasizing holistic integration of mind and body in action through occupational performance and the relationship between this holistic functioning, good health, and sense of well-being.

The call, spearheaded by Mary Reilly and supported by occupational therapy leaders such as Yerxa led to scholarly activity endeavoring to analyze and examine the state of knowledge of occupation. Subsequently, theory-based practice was enhanced by development of conceptual models of practice (referred to by Mosey [1970, 1996] as frames of reference). The attempt to understand theoretically the nature of occupation and its relation to health led Yerxa and colleagues at the University of Southern California to develop a doctoral degree program with a focus on occupational science in 1989 (Larson, 2006).

Definition and Scope of Occupational Science

When developed in the late 1980s and early 1990s, occupational science was conceptualized to be a basic scientific discipline, similar to sociology, biology, etc. (Clark, Parham, Carlson et al., 1991; Kielhofner, 2004; Larson, 2006). The founders of the new discipline defined it simply as "the study of humans as occupational beings" (Clark, Parham, Carlson et al., 1991, p. 300; see also Wilcock, 2003, p. 158). Specifically, Zemke and Clark (1996) stated that: "Occupational science is an academic discipline, the purpose of which is to generate knowledge about the *form, function*, and *meaning* of human occupation" (p. vii, emphasis mine). As understood in occupational science, occupation does not mean work. It refers to how humans occupy their time and includes self care, work/productivity (including paid and unpaid work such as house keeping, volunteering, child care, care for the elderly, etc.), and leisure pursuits (Law, Polatajko, Baptiste, & Townsend, 2002). In other words, occupation can be defined as "the ordinary and familiar things that people do every day..." (Clark, Parham, Carlson, et al., 1991, p. 300).

The founders observed that occupational science was not developing in the way a traditional science developed. It was not constrained by positivistic tradition of scientific inquiry (meaning testing of hypotheses through experimentation, relying exclusively on mathematical and statistical methods of analysis, and employing systematic, replicable procedures). Instead, the new discipline seemed to be more amenable to "subjective, qualitative approaches to inquiry...because of occupations' richness in symbolic meanings and the science's ethical roots in occupational therapy" (p. viii). That is not to say that such methods are not systematic, replicable, and open to public scrutiny like the experimental methods, because they are. The difference is that qualitative procedures are not based on experimentation, pursuit of detached objectivity, and use of mathematical and statistical methods of analysis.

Occupational science is now 16 years old. Since its founding in 1989, it has grown into a scientific movement embraced throughout the world. Occupational science symposia are held in different countries of the world. It has continued to emphasize its original interdisciplinary focus (Clark, 2006; Hocking, 2006). In 1993, the Journal of Occupational Science (JOS) was established to promote the study of humans as occupational beings. The scope of the new science has been a broad-based attempt to understand human occupation. This is evident in the topics that have been studied by occupational scientists as gleaned from literature published in the JOS.

Examples of such studies include: establishing a taxonomic understanding of the construct of occupation (Christiansen, 1994); investigating the biological basis of human occupations as indicated by evolutionary and anthropological evidence (Breines, 1995; Wilcock, 1993, 1995; Wood, 1998); exploring the use of objects in occupational performance as a source of identity (Hocking, 2000); occupational meaningfulness as a basin of attraction that structures assembly of occupational performance into patterns and determines the occupational life trajectory (Ikiugu, 2005); investigating the economic, social, and political significance of occupations such as crafts (Dickie & Frank, 1996); developing a theory of the human need for occupation based on basic human needs for adaptation, survival, capacity actualization, and the need to thrive (Wilcock, 1993); studying how individuals change their use of time (occupations) after traumatic brain injury (Winkler, Unsworth, & Sloan, 2005); exploring the theory of occupation of grooming as an outcome of the evolutionary process (Hartshorne, 2006), etc.

Research methods that have been used in occupational science include grounded theory methodology, phenomenology, ethnography, historical methods, time use diaries, Experiential Sampling Method (ESM), and so on. Assessing the viability of the discipline, Clark (2006) suggested that occupational science was strong, with occupation being a unifying paradigmatic focus, even though the researchers in the field had not come to a consensus as to how the construct was to be defined. She however expressed fear that unless occupational science really established itself outside of occupational therapy and became part of large, multidisciplinary teams, the discipline's continued survival would continue to be threatened.

Clark stated that apart from joining multidisciplinary scientific teams in centers such as the National Institute of Health (NIH), occupational scientists: "Not only must…encourage their colleagues in such centers to infuse an occupational perspective in their science, but they must guard against doing work that digresses from the occupational science paradigm" (pp. 171-172). In other words, occupational scientists must join multidisciplinary scientific teams and infuse occupational science in the discourses of those teams. In this book, an attempt is made to do precisely that. It is proposed that the occupational science perspective be infused in attempts to resolve pressing global and ecological issues that are necessarily of a multidisciplinary focus.

An Argument for Occupational Science as a Source of Conceptual Framework to Provide a New Understanding to Help Resolve Pertinent Global Issues

The basic premise of this book is that human occupational behavior has global consequences that affect the entire planetary system. If individuals understand these far reaching consequences of their occupational behavior, they may be willing to change it in such a way that the planet and its ecological system are affected positively. In other words, humanity created most of the global problems we face today through their daily occupational pursuits and related thinking processes and technological development. Consequently such global problems can be resolved largely through change in thinking processes and subsequent occupational performance patterns.

Understanding of this coupling between occupational performance and global issues of concern can be illuminated by knowledge derived from occupational science. The discipline can inform our understanding of how and why individuals choose the occupations in which they participate (Kilehofner, 2002; Law, Baptiste, Carswell, et al., 2000; Law, Polatajko, Baptiste, & Townsend, 2002), contextual factors that impact occupational performance (Christiansen & Baum, 2005; Dunn, Brown, & Youngstrom,

2003; Ikiugu, 2005; Ikiugu & Rosso, 2005), how occupational form as an organized, structured entity interacts with individual developmental structure to imbue occupational performance with meaning (Nelson & Jepson-Thomas, 2003), etc.

Such understanding can help us comprehend why individuals choose occupational performance patterns that may not be consistent with their own well-being or the health of the earth's ecology. Why does an individual choose to drive an 8 cylinder, 4.0 liter engine car that spews a large quantity of contaminants in the environment when driving a 4 cylinder 1.5 liter engine or even riding a bicycle would accomplish the same objective; to get from one point to another? Perception of the occupational form of driving according to the individual's developmental structure, cultural meanings of the occupational form, etc. probably determine the kind of car a person chooses to drive more than the simple objective of getting from point A to point B.

Understanding how and why individuals choose certain means of transportation rather than others could help us guide them so that they can find other ways of obtaining the same benefits (e.g. social prestige) without those means of transportation. Thus, for example, we can help people choose alternative ways of participating in the occupation of transportation that: 1) may be less extravagant (saving resources that can be used to address issues of poverty); and 2) contaminate the environment less. Understanding why and how individuals choose occupations and occupational performance patterns may help us develop guidelines to help them change their choices so that their performance is consistent with not only their own well-being but the well-being of all humans on earth and with global ecological health. Occupational science can provide knowledge necessary for such insights.

This is not the first time that it has been proposed that occupational science be used to provide guidelines to help modify human occupational performance for the benefit of the earth's ecological system. do Rozario (1997) argued that occupational scientists needed to be involved in facilitating occupational lifestyles consistent with the idea of deep ecology. She observed that this was necessary because we were living at a time when we were confronted with unprecedented challenges to human survival and sustainability of development. She postulated that: "Highly industrialized societies as well as rural and tribal communities are being faced with this same dilemma, namely the preservation of life and lifestyle, and of its sustainable development" (p. 112). She thus argued that we needed to shift our perspective in both ideology and actions both as organizations and as individuals because we could not continue to act in such a way that we "increasingly poison our means of breathing, eating, and surviving on this planet".

In essence, do Rozario concurred with Lovelock's (2006) argument that we could not continue treating the world as if it was our own private property which we could trash in any way we wanted. If we did that, it would be at our own peril. Furthermore, do Rozario postulated that we were connected with everybody and everything else [a view increasingly held by many leading scientists and philosophers (see Capra, 1996)]. Therefore, anything we did affected other people and other things, sometimes far away from us. More specifically, she stated that, "what we do on one side of the globe, country or neighborhood does affect what happens to the lives of people on the other side" (p. 112). We saw this truism in chapter 3, where it was pointed out that global warming accelerated by pollution caused mostly by activities in industrialized nations affected most severely the poor people living around the tropics. As an example, thousands of people in Kenya who were safe from malaria even two decades ago now die from the disease due to changed breeding patterns of mosquitoes caused by the raised temperatures.

It is clear that as a discipline which was created to facilitate the study and increased scientific understanding of human occupations and occupational behavior, occupational science has a responsibility to: illuminate the relationship between human occupational behavior and global problems that afflict all of humanity and our entire planet on the one hand; and how to change such behavior to increase chances of survival not only of humans but of the entire earth's ecosystem on the other. If as occupational scientists we shirk away from this responsibility, we repudiate one of the most important reasons for the existence of occupational science as a discipline. The work presented in this book is just a tiny step towards meeting that responsibility.

Reflection Exercise #4

After reading chapter 4, answer the following questions:

1. Before reading chapter 4, how aware were you (on a scale of one to ten with one being not at all aware and ten being completely aware) about the fact that:

   a. There is a new scientific discipline known as occupational science whose mandate is to generate knowledge about human occupation?

   b. Occupation refers not only to the job a person does but to things that a person wants, needs, or is expected to do, and can be categorized into productivity (work), self-maintenance, and leisure pursuits?

   c. The way you choose and perform occupations has an effect on issues that pose problems in the world as discussed in chapters one to three?

2. What can you do to learn more about occupational science and how your occupational performance affects the world?

3. How can you change your occupational performance so as to influence the issues discussed in chapters one to three positively?

References

Baum, C. M., & Christiansen, C. H. (2005). Person-environment-occupation-performance: An occupation-based framework for practice. In C. H. Christiansen, C. M. Baum, & J. Bass-Haugen (Eds.), *Occupational Therapy: Performance, Participation, and well-being* (3rd ed.). Thorofare, NJ: Slack.

Breines, E. (1995). *Occupational therapy activities: From clay to computers*. Philadelphia: F.A. Davis.

Capra, F. (1996). *A new scientific understanding of living systems: The web of life*. New York: Anchor Books/Doubleday.

Christiansen, C. (1994). Classification and study in occupation: A review and discussion of taxonomies. *Journal of Occupational Science*: Australia, 1(3), 3-20.

Clark, F. (2006). One person's thoughts on the future of occupational science. *Journal of Occupational Science, 13*(3), 167-179.

Clark, F. A., Parham, D., Carlson, M. E., Frank, G., Jackson, J., Pierce, D., et al. (1991). Occupational science: Academic innovation in the service of occupational therapy's future. *American Journal of Occupational Therapy*, 45, 300-

Dickie, V. A., & Frank, G. (1996). Artisan occupations in the global economy: A conceptual framework. *Journal of Occupational Science*: Australia, 3(2), 45-55.

Dunn, W., Brown, C., & Youngstrom, M. J. (2003). Ecological model of occupation. In P. Kramer, J. Hinojosa, & C. B. Royeen (Eds.), *Perspectives in human occupation: Participation in life* (pp. 222-263). New York: Lippincott Williams & Wilkins.

Dunton, W. R. (1919). *Reconstruction therapy*. Philadelphia: W.B. Saunders.

Hartshorne, S. (2006). An evolutionary perspective of grooming as an occupation. *Journal of Occupational Science, 13*(2), 126-133.

Heemskerk, M., Wilson, K., & Pavao-Zuckerman, M. (2003). Conceptual models as tools for communication across disciplines [Electronic version]. *Conservation Ecology, 7*(3), 8-20. Retrieved December 11, 2006, from http://www.consecol.org/vol7/iss3/art8.

Hocking, C. (2006). *Editorial. Journal of Occupational Science, 13*(3), 166.

Hocking, C. (2000). Occupational science: A stock take of accumulated insights. *Journal of Occupational Science, 7*(2), 58-67.

Ikiugu, M. N. (2007). *Psychosocial conceptual practice models in occupational therapy: Building adaptive capability.* St Louis, MO: Elsevier/ Mosby.

Ikiugu, M. N. (2005). Meaningfulness of occupations as an occupational-life-trajectory attractor. *Journal of Occupational Science,* 12, 102-109.

Ikiugu, M. N., & Rosso, H. M. (2005). Understanding the occupational human being as a complex, dynamical, adaptive system. *Occupational Therapy in Health Care, 19*(4), 43-65.

Kielhofner, G. (2004). *Conceptual foundations of occupational therapy* (3rd ed.). Philadelphia: F.A. Davis.

Kielhofner, G. (2002). Volition. In G. Kilehofner (Ed.), *Model of human occupation: Theory and application* (pp. 44-62). Philadelphia: Lippincott Williams & Wilkins.

Larson, E. (2006). Occupational science. *Research Library: University of Wisconsin Board of Regents.* Retrieved December 11, 2006, from http:// www.education.wisc.edu/occupational_science/researchlibrary/#history.

Law, M., Baptiste, S., Carswell, A., McColl, M. A., Polatajko, H., & Pollock, N. (2000). *Canadian Occupational Performance Measure.* Ottawa, ON: CAOT Publications ACE.

Law, M., Polatajko, H., Baptiste, S., & Townsend, E. (2002). Core concepts of occupational therapy. In E. Townsend (Ed.), *Enabling occupation: An occupational therapy perspective* (pp. 29-56). Ottawa, ON: Canadian Association of Occupational Therapists.

Lovelock, J. (2006). *The revenge of Gaia: Earth's climate crisis & the fate of humanity.* New York: Basic Books.

Mosey, A. C. (1996). *Psychosocial components of occupational therapy.* New York: Lippincott Williams & Wilkins.

Mosey, A. C. (1970). *Three frames of reference for mental health.* Thorofare, NJ: Slack.

Nelson, D. L., & Jepson-Thomas, J. (2003). Occupational form, occupational performance, and a conceptual framework for therapeutic occupation. In P. Kramer, J. Hinojosa, & C. B. Royeen (Eds.), *Perspectives in human occupation* (pp. 87-155). New York: Lippincott Williams & Wilkins.

Quiroga, V. A. (1995). *Occupational therapy: The first 30 years - 1900 to 1930.* Bethesda, MD: American Occupational Therapy Association.

do Rozario, L. (1997). Shifting paradigms: The transpersonal dimensions of ecology and occupation. *Journal of Occupational Science, 4*(3), 112-118.

Townsend, E. (1997). Occupation: Potential for personal and social transformation. *Journal of Occupational Science: Australia, 4*(1), 18-26.

Wicks, A. (2005). Understanding occupational potential. *Journal of Occupational Science, 12*(3), 130-139.

Wicks, A. (2000). International developments in occupational science. *Journal of Occupational Science, 7*(2), 87-89.

Wilcock, A. A. (2007). Occupation and health: Are they one and the same? *Journal of Occupational Science, 4*(1),

Wilcock, A. (2006). *An occupational perspective of health (2nd ed.).* Thorofare, NJ: Slack.

Wilcock, A. A. (2003). A science of occupation: Ancient or modern? *Journal of Occupational Science, 10*(3), 115-119.

Wilcock, A. A. (2001). Occupation for health: Re-activating the regimen sanitatis. *Journal of Occupational Science, 8*(3), 20-24.

Wilcock, A. (1995). The occupational brain: A theory of human nature. *Journal of occupational science: Australia, 2*(1), 68-72.

Wilcock, A. (1993). A theory of the human need for occupation. *Occupational Science: Australia, 1*(1), 17-24.

Winkler, D., Unsworth, C., & Sloan, S. (2005). Time use following a severe traumatic brain injury. *Journal of Occupational Science, 12*(2), 69-81.

Wood, W. (1998). Biological requirements for occupation in primates: An exploratory study and theoretical analysis. *Journal of Occupational Science, 5*(2), 66-81.

Zemke, R., & Clark, F. (1996). Preface. In R. Zemke & F. Clark (Eds.), *Occupational science: The evolving discipline (pp. vii-xviii).* Philadelphia: F.A. Davis.

Chapter 5
Human Occupational Performance
as a Partial Cause of Global Problems

Think about the following scenario: You wake up in the morning; brush your teeth; take a shower; get dressed; may be make some coffee; probably grab a little breakfast with your coffee; get in your car and drive to work; do work related activities; probably take a coffee and lunch-break sometimes during the day; drive back home after a long day's work; may be take another shower and change clothes; make dinner and eat with your family or go out to dinner with your family; or may be go on a date; watch some television or perhaps spend a romantic evening with a loved one; and finally go to sleep. This is a typical day for many working adults all over the world. The question is, what would you say if someone told you that whether or not and/or the way you perform any of the above enumerated activities (known as occupations in occupational science) every day affects the following issues positively or negatively: the prevalence and consequences of poverty in the world; material inequalities either within your country of residence or between nations; global warming and climate change, and related consequences; corruption and institutional failure; over-population; etc?

The core argument advanced in this book is that there is a relationship between things that human beings do as they pursue daily occupations as described in the above scenario and many of the problems that we face in our world today. This assertion has been supported by Wilcock (2006) who argued that the occupational nature of humans was partly to blame for current environmental and ecological concerns and loss of essential community values. One could even assert that there is a causal relationship between occupational performance and the various problem variables. However, because of the complexity of the earth and its eco-system, such a causal relationship cannot be easily and reliably ascertained. Never the less, it has been demonstrated in the reviewed literature in earlier chapters that human occupational behavior contributes to such problems.

It is important to remember that even though we have so far focused on the negative consequences of human occupational behavior, there are also positive outcomes. Through pursuit of daily occupations, humans individually and collectively have been able to contribute to establishment of a better, more comfortable life for all people. As an example, discovery of increasingly effective medications by those who have devoted their lives to the art of healing has led to longer and better quality of life for many people. As will become apparent later in the book, there are individuals who are also devoting their lives to poverty alleviation. Even some have chosen to engage in occupational behavior devoted to saving the planet and making it more beautiful and habitable for all life through activities such as re-a forestation.

In this chapter, the postulated global consequences of human occupational behavior will be examined in greater detail using a conceptual model (see figure 5-1 on opposite page). Examples from literature and personal experiences will be used to explain and illustrate the model.

Figure 5-1 depicts a conceptual model constructed following the examples published by Heemskert, Wilson, and Pavao-Zuckerman (2003). Heemskert et al. argued that effective ecosystem management could only be achieved through interdisciplinary efforts where both ecological and social scientists collaborated to solve the planet's problems. However, they observed that this interdisciplinary collaboration was hindered by lack of a common language that was understandable across disciplines among other things. They set out to develop conceptual models that could be used to communicate across disciplines and to facilitate such collaboration.

The modeling by Heemskert et al. was done in a workshop involving 26 participants and 2 facilitators from 11 universities throughout the United States, representing four Integrative Graduate Education Research and Training (IGERT) and six Long-Term Ecological Research (LTER) programs. Their models were constructed using symbols derived from Odum (1983) that were commonly employed by ecologists. However, use of Odum's symbolism was found to be too complicated and confusing for the purpose of the conceptual model proposed in this book. Instead, the conceptual mapping method was adopted where arrows indicating interaction between variables depicted in the model were used. The interactions were explained and illustrated with evidence derived from literature or personal experiences. As can be seen in figure 5-1, the human being was depicted as the agent driving the global variables of poverty, global warming, material inequalities, diseases, institutional dysfunctions, and population growth through occupational performance in the areas of productivity, leisure pursuits, self-maintenance, sex, and war.

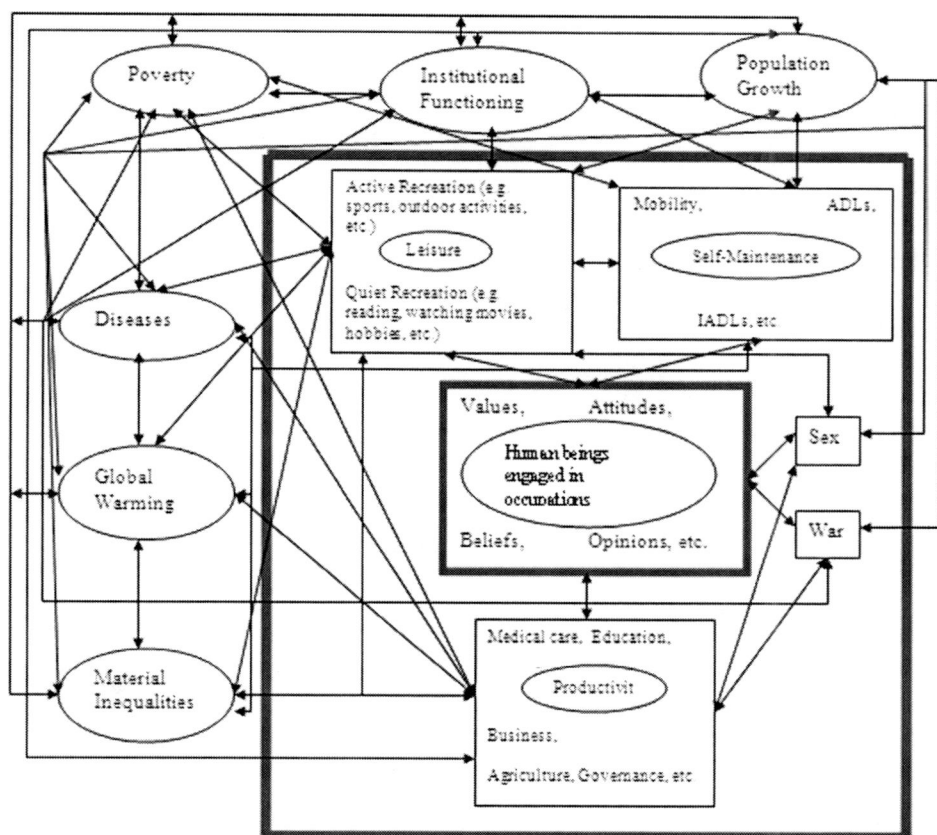

Figure 5-1. A conceptual model depicting the consequences of human occupational performance in the categories of productivity, leisure, self-maintenance, sex, and war that impact global problems such as poverty, global warming, diseases, material inequalities, institutional dysfunction, and population growth, and the interactions between all the variables.

Human Being: The Center of Agency

Figure 5-1 presents the human being and his/her occupational performance and its consequences as a complex system. At the center of this system is the human being, whose actions are determined by the perceived meaning of performance, based on the person's goals, values, and experiences in life (Baum & Christiansen, 2005; Ikiugu, 2005; Ikiugu & Rosso, 2005) in addition to the need to adapt and survive (Christiansen, 1994). This meaning emanates from intrinsic factors comprising of the values, attitudes, beliefs, opinions, etc. that a person holds. Attitudes may be defined as a person's general beliefs and opinions about other people, things, situations, and so on (Harrow, 1996). One can have different, sometimes even ambivalent attitudes about something. For example, a person may believe that studying is an enjoyable activity while at the same time holding the opinion that it is frustrating and stressful.

In essence, attitudes and motives may be seen as being rooted in feelings (Personal Growth Center, n.d.). Values, opinions, beliefs, and sentiments on the other hand are mental views about phenomena. Values help us resolve contradictions emanating from attitudes, thus guiding our process of prioritizing and making choices. They tend to be more deep-seated than attitudes and beliefs because they are "rooted to various institutions and processes in society..." (Gecas & Libby, 1976, p. 37). We assign value to phenomena based on our education, family experiences, culture, religious affiliations, and so on.

We can argue that the way a person chooses and performs occupations in the areas of productivity, leisure pursuits, and self-maintenance is related to his/her attitudes, opinions, and beliefs about those occupations, and his/her valuation of them. In turn, valuation of occupations gives them meaning. The meaning of an occupation and therefore the motivation to engage in it may be seen as "socially and culturally influenced but becomes internal to the individual through personal interpretation" (Baum & Christiansen, 2005, p. 248).

Values ("what one finds important and meaningful to do" [Kielhofner, 2002, p. 15]); attitudes (psychological tendencies that are "expressed by evaluating a particular entity with some degree of favor or disfavor" [Cozzarelli, Wilkinson, & Tagler, 2001, p. 208]); beliefs that are the basis of those values and attitudes; and opinions resulting from those values, attitudes, and beliefs, result from a person's experiences as he/she interacts with both the human and non-human environmental context, and his/her interpretation of those experiences. Furthermore, a person's values, attitudes, beliefs, opinions, and so on, that give meaning to occupations result from his/her experiences as he/she interacts with the environment. This meaning in turn provides the motivation that determines a person's choice of occupations in which he/she participates.

Based on the above intrinsic factors, the human being chooses and participates in occupations. From the perspective of occupational science and occupational therapy, such occupations are categorized into three areas: self-maintenance, productivity, and leisure (American Occupational Therapy Association [AOTA], 2002; Hocking, 2000; Law, Polatajko, Baptiste, & Townsend, 2002). Several variations of the above classification exist in occupational therapy and occupational science literature. Some of these categorical variations include: work, play, rest and sleep; school/work, play/leisure/recreation, Activities of Daily Living (ADLS), and family interaction occupations; work, play, and ADLs; productivity, leisure, and self care; work, play, and self maintenance; and work, leisure, and ADLs (Christiansen, 1994).

In the above categorization, it seems that in the perspective of occupational therapy and occupational science, occupational classification may be conceptualized as consisting of: productive endeavors, including paid work, volunteering, homemaking, care for others, and so on (in this book, I also include medical care, education, business, agriculture, and governance in this category because those occupational endeavors are crucial to our understanding of the relationship between productivity and global issues under consideration); taking care of oneself, including obtaining nutrition, dressing, toileting, bathing, taking care of personal hygiene (all referred to as Activities of Daily Living [ADLs]), community mobility including use of public transportation, doing laundry, grocery shopping, and so on (referred to as Instrumental Activities of Daily Living [IADLs]); and leisure pursuits including quiet recreational activities such as reading, engaging in hobbies, socialization, and so on, and active recreation including playing sports, hiking, and other physically demanding activities (American Occupational Therapy Association [AOTA], 2002; Law et al., 2002).

Sexual activity and war are presented separately because in my view, they do not fit neatly into any of the above discussed occupational performance categories. For example, for those holding the view that sexual activity is a biological necessity similar to eating, breathing, or regulating body temperature (Gecas & Libby, 1976), sex may be classified as a self-maintenance occupation. For others who hold the recreational philosophy of sex, sexual intercourse may be seen merely as recreational and a source of pleasure. Similarly, the purpose of war is largely killing and destruction, and therefore it does not fit into any of the three categories of occupations as defined by occupational therapists and occupational scientists. Sometimes useful technology results from endeavors to develop war instruments. Good examples of these technological developments that have revolutionalized human civilization include invention of the cell phone, air travel, etc. Never the less, these positive outcomes are not the objective of war itself. They are lucky positive side effects of this deliberately destructive occupational pursuit.

The theses in this book is that human choices and occupational performance in the above discussed occupational performance areas have significant global consequences, both positive and negative, which include impacting material inequalities, global warming/climate change, diseases, poverty, institutional functioning, and population growth. Those outcomes in turn have an effect, through feedback mechanisms, on how human beings choose and participate in the various occupational performance areas. This complex relationship will be discussed in more detail.

Global Consequences of Individual Productive Occupational Behavior

Performance of productivity-related occupations both at the individual and collective levels affects many of the global issues highlighted in this book. At the individual level, a person's attitudes, beliefs, values, and opinions determine his/her choice of occupations and subsequent occupational performance patterns (Ashford, 2001). Many of these attitudes, values, behaviors, and beliefs are rooted in the person's culture and through their influence on performance actually determine the person's upward or downward mobility at the job, and subsequently his/her economic status. Individual performance has a direct effect on poverty and material inequalities because it determines, to a certain extent, the individual's economic status. I say to a certain extent because contrary to what some people may think (that individual failure is purely attributable to personal shortcomings), there are those who may have the right attitudes and aptitudes, and work very hard, but structural barriers inherent within social systems keep them poor.

Individual performance also affects issues at the macro-level. Activities engaged in by a person that lead to poverty contribute to material inequalities which have been suggested to have deleterious effects on economic growth of nations (Fuentes, 2005; Goransson, 2004). For one, poverty causes conflicts which affect productivity negatively. A friend of mine recently told me about his recent visit in Kenya. During his visit, he observed that it was difficulty nowadays to enjoy the nice things that one may have in the country, because many people who do not have those things are tempted to steal them. He mentioned to me the trend where the more educated armed gangsters hold people up at gunpoint and take their vehicles. Sometimes they hijack public transportation vehicles, commandeer them to automatic teller machines, and force passengers to withdraw money from their accounts and give it to them. These types of crimes are largely a direct consequence of extreme poverty, lack of jobs even for educated people who are willing to work, and gross inequalities in society, in addition to an inefficient law enforcement system.

Under these circumstances, individuals with capital may be reluctant to live and invest in the country, which affects economic growth adversely. In addition, according to Fuentes (2005), Goransson (2004), and Salai-i-Martin (2002), inequalities and poverty tend to slow down economic growth. This is because poor people are not able to participate in economic discourse effectively. They do not have the money to buy goods, which affects manufacture of those goods because of lack of a ready market. Poverty also tends to be associated with poor health (see discussion in chapter one), which means that poor people are less able to work and be productive because of illness.

Some individuals also hold positions that place them in a vantage point to be able to impact issues at the macro level more directly through their occupational performance. Social scientists point out that three occupational performance problems that adversely affect macro-economics are bureaucratic corruption, inefficiency, and incompetence (Mauro, 1998; Mbaku, 1991, 1996; Kimenyi, 1987). Bureaucratic corruption is particularly significant because it involves illicitly paying bureaucrats in order to realize excessively high profits, a behavior known as "rent-seeking", which has disastrous effects on competition and therefore on productivity and quality of goods and services. It is also related to bureaucratic occupational behavior of regulating economic activities, which in turn is related to political occupational behavior of governance and decision-making.

Through rent-seeking activities inherent in corrupt systems, politicians and civil servants are influenced to make public policies that result in patterns of wealth distribution within and between nations designed to favor specific interest groups. Examples of this effect abound. Readers may recall the controversy of campaign contributions by interest groups, most of them from rich multinational companies, a prevalent practice in the conservative controlled US congress, senate, and administration in the early 2000s. According to a survey conducted by Anderson, Cavanagh, Hatman, Klinger, and Chan (2004) on behalf of the Institute for Policy Studies (IPS), 50 companies contributed about US$10.4 million to various candidates during the 2004 election cycle.

Incidentally, these same 50 companies led in outsourcing jobs to places like India where executives and workers could be paid only a fraction of what their American counterparts would earn. This led to loss of millions of jobs for American workers and congress did nothing to stem the trend. Instead, they legislated to give these companies hefty tax-breaks and the administration tried to convince the American public that outsourcing is actually good for American workers, with Bush's top economic adviser telling congress that "outsourcing was a 'good thing'" (p. 4) and the Treasury secretary expressing similar views. The administration's argument was that outsourcing lowers labor costs and boosts profits leading to creation of more jobs for the American workers[iv].

According to the findings by Anderson et al., while it is true that outsourcing lowers labor costs and increases profit margins, the increased profits do not go into creation of more jobs but rather into increased remuneration of top company executives irrespective of the actual performance of the companies they lead. Forty one executives who contributed the most money to campaigns in the 2003-2004 elections earned an average of US$17.4 million in 2003 which was more than twice the normal average earnings of the executives of large companies. Furthermore: "CEOs of the 69 companies that sponsored this summer's political conventions (referring to summer 2004) averaged $9.2 million in total compensation in 2003, representing a hefty 52 percent raise over their average pay in 2002" (p. 1).

As the earnings of CEOs who made substantial campaign contributions to federal legislative candidates increased, the gap in pay between CEOs and regular workers also increased. The average worker to CEO pay ratio increased from 1:282 in 2002 to 1:301 in 2003. According to Anderson et al., if the workers' pay had increased at the same rate as CEO pay, the minimum wage would have been $15.76 per hour in 2004. In this case, campaign contributions by CEOs to federal legislative candidates acted as bribes to encourage legislators to give companies tax-breaks and look the other way as the companies outsourced jobs leading to unemployment of millions of American workers. This was a kind of 'rent-seeking' behavior.

The occupational activities of executives and legislators in this case led to redistribution of wealth with executives earning many more times the average workers and this widened the gap between the rich and the poor in the country. Such decisions and subsequent occupational pursuits of individuals in key public positions may be to blame for the increasing inequalities in wealth, and gradual sliding of the middle class into poverty. The subsequent result is that currently, there are those who are rich and then there is everybody else. This has led to the 'swallowing' of the American dream (Knox, 2006). According to the report by Knox, some see this trend as a recipe for future social conflict and upheaval in the United States.

The above occupational behavior that creates increased poverty and a widening gap between the poor and the rich is not unique to the USA. Similar trends have been reported in Australia and Kenya among other countries (Australian Council of Super Investors Inc., 2006; Goransson, 2004). In Kenya, corruption affects occupational behavior of public officials more directly with even more ominous effect on poor people. Such pernicious occupational behavior of dishonest business people and public officials was evident in the Anglo-Leasing scandal revealed in 2004 (Githongo, 2005; Vasagar, 2005).

The Anglo-Leasing scandal (see earlier reference in chapter one) was an example of direct looting of public funds by public officials in collaboration with corrupt business people. In 2002, the Republic of Kenya sought to modernize the passport printing equipment. The government embarked on the process of outsourcing the acquisition of such equipment and the Forensic Science Laboratories for the police force. A French company gave a bid to do the job for 6 million Euros. However, the government gave the contract to a British firm (Anglo-Leasing Company) for 30 million Euros, which was 5 times higher than the quote given by the French firm to which Anglo-Leasing was to sub-contract any way. As it turned out, Anglo-Leasing was a fictitious company connected to a Ms. Sudha Ruparell, daughter of Chamanlal Kamani and a sister to Rashmikant Chamanlal Kamani and Deepak Kamani, a politically well connected family in Kenya that had been involved in a number of scandals in the past.

The contract between the Kenya Government and Anglo-Leasing company was never announced for public bidding, and the fictitious company was paid a 3% down payment amounting to Ksh90 million (Githongo, 2005). The shady deal among others was revealed by Sir Edward Clay, the then British High Commissioner to Kenya (Vasagar, 2005). This deal was obviously designed by an individual business person (Sudha Ruparell either alone or in collaboration with others) and endorsed by corrupt Kenyan officials in pursuit of occupations pertaining to their official positions.

According to the investigation by Githongo, the former Kenya permanent secretary for ethics, implicated officials included Mr. Chris Murungaru (the then minister in charge of security), Moody Awori (Vice-President), Kiraitu Murungi (then justice minister), and David Mwiraria (the then finance minister). Mr. Murungi and Mr. Mwiraria in particular tried to pressure Mr. Githongo to drop the investigation into the scandal by promising that some loans that his father owed would be forgiven if he did so. In this case, occupational pursuits by individuals acting in their official capacity led to a conspiracy to defraud the Kenyan government millions of Kenya Shillings that could be used to improve the welfare of the more than 50% of the Kenyan population living in abject poverty (on less than a dollar a day [see discussion in chapter one]). Such deals, according to the Transparency International (TI), have led to loss of at least KSh85 billion a year, which would be enough to fund free primary education for at least 10 years (Namunane, 2006).

Through their occupational behavior, individuals in those deals contributed significantly to increasing poverty and the ever-widening gap between the rich and the poor in the country. Notice that in this case, Githongo demonstrated appropriate occupational behavior for public officials in pursuit of their occupations in the productivity category. He diligently pursued investigations (an occupational task associated with his office) even under pressure from powerful corrupt public officials to stop, to the point of losing his job and having to live in exile in fear for his life. This is an example of how individuals can elect to perform occupations in the productivity category for the good of the public, impacting global issues positively. Through his occupational behavior, along with actions by Sir Edward Clay, a devious plan that could have resulted in loss of more billions of Kenya Shillings and a significant increase in the level of poverty in the country was averted.

Another way in which occupational behavior in the area of productivity is related to global issues of concern is through its direct impact on environmental change and global warming/climate change. As discussed in chapter 3, productive occupational behavior may and often leads to destruction of the earth's vegetation cover (e.g. agricultural activities) and production of pollutants leading to environmental pollution and global warming. This is even more so in modern times because as Gore (2006) aptly observed in his movie, the shovel had become larger with mechanization making it possible for an individual to affect the land at a larger scale, instruments of war no longer had the potential to cause just soldier casualties but could now wipe out entire cities, etc. In this highly mechanized world, agricultural activity led to unprecedented destruction of natural forests and vegetation cover with the subsequent risk of desertification.

Industrial and transportation activities on the other hand led to accumulation of greenhouse gasses in the atmosphere adding substantially to the problem of global warming and climate change. Recently, it was suggested that airline flights have increased significantly over the last few years and air transportation is expected to continue to skyrocket (Stoller, 2006). But aircraft produce a considerable amount of greenhouse gasses. According to scientists, a commercial jet from New York to Denver produces 840 to 1660 pounds of carbon dioxide per passenger, an amount of carbon dioxide generated by a Sports Utility Vehicle (SUV) for one month. That is why it is proposed that by the year 2050, commercial airlines could be the greatest contributor to global warming.

The consequences of climate change and global warming continue to come to light every day. Recently there was a report that a huge chunk of the Greenland ice sheet had broken off and was floating out to sea. This could pose considerable threats to the shipping industry. This breaking off of the ice-sheet was attributed to global warming. Also, a new report stated that the polar bears were now being threatened by the global warming and subsequent melting of the polar ice (Davis, 2006). In addition, it had been predicted that 2007 would be the hottest year in human history [(Satter, 2007), a prediction largely coming true during this writing]. This was in part due to a cyclical warming trend known as El Nino, combined with the global warming due to greenhouse gasses.

Thus, the warming process due to pollution has a multiplier effect on the natural warming trends. When as an individual one decides to book a flight or to drive an SUV in the process of conducting personal business, responsible, reflective occupational performance may mean asking himself/herself whether this is the only way the task can be accomplished. May be the occupational task can be completed in another way that does not involve contributing to accumulation of greenhouse gasses. Could the task for which one wishes to travel be accomplished through fax or email communication rather than one having to drive or fly? If one has to travel to another point, could he/she drive a smaller car rather than a big SUV that consumes much gasoline and spews so much carbon dioxide into the atmosphere?

Executives of larger companies could make the greatest difference if they commit to change. Japanese car makers Honda and Toyota, realizing the gravity of global warming, decided to build hybrid passenger vehicles whose motors run on both electrical and fossil fuel energy. This has led to significant changes in fuel consumption with some of the vehicles attaining up to 40 miles per gallon of fuel. If all companies followed the example of Toyota and Honda, eventually, hybrid cars would be affordable to the general population and this would contribute significantly to cutting down on gasoline consumption. Also, why are other more environment-friendly fuels not being developed? For example, hydrogen fuel is used in rockets that propel space vehicles. Why can't this form of fuel be developed for general use by aircraft and even cars? If tax cuts and subsidies being extended to the oil industry by the US congress and administration had been used to encourage executives in those companies to develop alternative fuels, it is possible we could already have been far along in solving the problem of environmental pollution due to travel and transportation business.

One may think that such proposed reforms are nothing but utopian dreams. However, there are examples of individual executives who have made a difference by committing to doing things differently. A good example is a Mr. Ray Anderson, Chairman of the textile manufacturing company known as *Interface* in Georgia (Baer, Bluestein, Chafkin, et al., 2006). Interface made carpet tiles used in the floors of airport lounges, office buildings, and even residential houses. These tiles were made mostly from petrochemical materials. Many of the customers of the company were architects, interior designers, etc., and many of them tended to be environmentally conscious. They started asking questions about what the company was doing for the environment. This forced Mr. Anderson to inquire from his engineers about what it took to make the products that the company sold.

Mr. Anderson was astonished to find that, according to his engineers, the company used 1.2 billion pounds of raw materials per year, most of them oil and natural gas related, and the waste from those raw materials was ultimately incinerated. This finding was shocking to Anderson who stated that he was astonished and wanted to "throw up" because his "company's technologies and those of every other company" that he knew of "anywhere, in their present forms, are plundering the earth" (p. 80). He realized that those costs that are referred to in business lingo as "externalities" because they do not figure anywhere in a company's analysis of cost of production and therefore do not affect the bottom line, in effect are extremely expensive to humankind in general. He realized that he, and other companies similar to Interface, were in his own words "thieves" who were robbing the earth and all its inhabitants in the most dangerous manner. This epiphany led to institution of measures to change the way Interface did business so that sustainability became a fundamental company goal.

Although Lovelock (2006) did not believe that sustainable development was possible at this point in our planet's history (see chapter 3), the changes made by Interface, led by Anderson, resulted in waste reduction through recycling, development of products that did not require certain types of noxious raw materials, e.g. carpets that were self-adhesive and did not require use of glue, energy usage reduction, etc. Such reforms were not only environmentally friendly but in fact led to significant savings for the company. Interface saved "more than $60 million in the first three years. To date, cost reductions for waste have amounted to more than $300 million, enough to finance Interface's research into the heart of its sustainability problem..." (p. 80).

The case of Interface is an example of how an individual executive's decision, in his pursuit of occupations in the area of productivity, may have a significant impact on global issues such as environmental destruction and global warming/climate change. Anderson's actions in guiding Interface went counter to the generally held myth that profitable business was inconsistent with taking responsibility of global issues such as poverty, diseases, material inequalities, environmental destruction and global warming/climate change, etc. This myth originated from Adam Scott Smith's (1776) doctrine of self-interest which became the gospel of adherents of the free-market business system. This doctrine glorified greed and selfishness and portrayed generosity, egalitarianism, and demonstration of caring as un-natural (even immoral) and a weakness that needed to be somehow punished. This attitude was clear in the prevalent attitude towards the poor in the US and probably other capitalist nations as demonstrated in the discussion latter in this chapter.

The problem, however, is that this type of irresponsible self-interest is different from what Mosey (1970, 1996) called enlightened self-interest. In enlightened self-interest, the individual realizes that meeting his/her needs is dependent on the needs of others being met. We can even argue that enlightened self-interest means realizing that meeting one's needs is dependent upon the needs of other life-forms and the planetary ecological system in general, not only humans, being met. As Obama (personal communications, 2007 April 1st) asserted, we are our brother's keeper after all. Injustice or suffering anywhere in the remotest parts of the world is injustice and suffering to us all. This view was expressed by Capra and many mystical thinkers before him such as Chief Seattle, an indigenous American, who once stated that:

> This we know.
> All things are connected
> like the blood
> which unites one family...
> Whatever befalls the earth,
> befalls the sons and daughters of the earth.
> Man did not weave the web of life;
> he is merely a strand in it.
> Whatever he does to the web,
> he does to himself. (Cited in Capra, 1996, p. xi)

Anderson realized this and took action accordingly. The validity of this view is apparent in the success that Interface subsequently experienced.

Political participation is another occupation that is not very well delineated in the occupational therapy or occupational science literature. In this book, it is argued that it is an occupation which may be classified under the category of productivity. Tasks that constitute the occupation of political participation include not only voting and legislation of laws by those who are elected to represent the people but also working in campaigns, contributing financially, contacting elected officials, lobbying, attending political rallies, participating in political protests, participating in town hall meetings, and other forms of political engagement tasks (Brady, Verba, & Schlozman, 1995; Verba, Schlozman, & Brady, 1995). It is also important to point out that political participation is not only a productive occupation in it's own right as conceptualized in this book, but it significantly impacts other occupations in all areas of performance. Political occupational pursuits determine decision making in governance and subsequently in distribution of wealth and therefore influence the extent of material inequalities in society, whether or not coordinated efforts are made to curb environmental destruction and global warming/climate change, etc.

We have already seen that the election of George W. Bush led to US withdrawal from the Kyoto treaty, which the previous Vice-President Al Gore had signed under the Clinton administration. Election of Bush into office and his subsequent decisions in pursuit of his occupation as the President of the United States had a significant impact on whether or not the US supported the treaty on global warming and subsequently, whether or not the treaty was likely to succeed considering that the US is the world's largest contributor to the accumulation of greenhouse gasses in the environment. We have also seen how participation in political occupational behavior through political campaign contributions by company executives led to legislation that contributed to the widening gap between the rich and the poor in the United States and Australia among other nations (ACSI, 2006; Anderson et al., 2004).

Another example of the interface between political occupational performance and global issues can be seen in legislations that have tended to weaken Federal and State programs designed to assist poor people financially, in effect further widening the gap between the poor and the rich. A good example of such legislation is the welfare reform law (Personal Responsibility and Work Opportunity Act, P.L. 104-193) enacted by Congress and signed by President Clinton in 1996 which did away with any obligation for the Federal or State government to financially assist, even minimally, those who were the most vulnerable in society, including helpless children from the poorest families.

Under this law, all kinds of social assistance programs such as Aid to Families with Dependent Children (AFDC), Emergency Assistance (EA), General Assistance (GA), Supplemental Security Assistance (SSA), and Food Stamps were significantly reduced. By Congress enacting this law, each of the congressmen or congresswomen who voted for the legislation, in pursuit of their occupational performance as legislators, in effect condemned millions of the most vulnerable individuals in society to even more suffering due to lack of food and other basic necessities required to keep body and soul together. By signing the legislation into law in his occupational performance as president, Clinton supported this Draconian affront against the most vulnerable members of the American society (Anelauskas, 2005).

There are many variables that determine how political occupational roles are performed. However, it may be reasonable to argue that how legislators enact laws in pursuit of their occupational endeavors is dependent on their perception and attitudes towards issues. For example, attitudes towards poverty and the poor determine whether or not legislators enact laws that are for or against the poor. According to social scientists, attitudes towards the poor (psychological tendencies expressed by evaluating the poor with some degree of favor or disfavor) are important because they "are likely to have significant consequences for poor people themselves, especially in terms of the impact of these attitudes on middle-class voting behavior, willingness to help alleviate or end poverty, and beliefs about welfare and welfare reform" (Cozzarelli, Wilkinson, & Tagler, 2001, p. 208).

Thus, such attitudes determine the kind of people individuals elect into office in the process of pursuing their political participation occupations, and subsequently, the kind of legislation that such elected representatives may enact and how such legislation could affect the poor. This is important considering that according to Cozzarelli et al., in 1998 there were about 34.5 million people (12.7% of the American population) who were below the federal poverty line in the USA. Today, that number is probably higher according to the latest poll by the Agence France-Presse (2007). It is important to understand general attitudes towards the poor in order to understand how pursuit of occupations in the area of political participation may affect material inequalities and poverty.

Attitudes towards the poor are expressed in attributions that people make towards poverty. There are three categories of such attributions (Cozzarelli et al., 2001; Hunt, 2002; Shek, 2004): 1) Individualistic (poverty is attributed to individual or internal factors. People who hold this view tend to blame the poor for their condition. They argue that people are poor because they lack intelligence, have no motivation, are lazy, are not stringent enough with regard to saving money, are wasteful, etc.). 2) Structuralist (poverty is attributed to socio/political barriers such as businesses and industries keeping wages low, industries not providing good jobs, prejudice, discrimination, and lack of opportunities for some people, e.g. the minorities, and so on). 3) Fatalistic (people are poor because of bad lack, such as illness, supernatural causes, etc.).

In the study conducted by Cozzarelli et al. (cited earlier) involving 209 undergraduate college students from a large Midwestern University, it was found that the participants were likely to hold internal or individualistic attributions to poverty. Incidentally, a significant number of those participants also tended to lean towards conservatism in political affiliation, and tended to believe in the American protestant ethic of a 'just world' doctrine. The protestant ethic is based on the premise that opportunity is equally available to all, and those who are wealthy tend to be so because of their own effort. Furthermore, the 'just world' view which is also associated with the protestant ethic is based on the belief that good things happen to those who deserve. Thus, if you are wealthy, it is because you deserve to be so, and if you are poor, you similarly got what you deserve. Subsequently, the politically conservative individuals, according to Cozzarelli et al. are also likely to hold the protestant ethic point of view (along with the 'just world' perspective) and therefore tend to blame the poor for their condition.

This attitude may explain why federal and state programs that were designed to assist the poor financially have been so dramatically reduced since the conservative republican party became the majority in both the Congress and Senate in the early 1990s, a situation that led Anelauskas (2005, p. 1 of 10) to assert that the: "Official American social policy on the national level has seemingly become: Poverty isn't the enemy, the poor are...As a nation, the United States of America seems to have taken to heart the bitter ironic humor of George Bernard Shaw, who once wrote, 'I hate the poor and look forward eagerly to their extermination.'"

Such attitudes may have even more far-reaching consequences. Those who believe in individualistic attribution of poverty may tend to view certain races of people as inherently (genetically) inferior and therefore as naturally fated to be poor. Brand (1997) argued that it is a truism that there are racial differences in intelligence, with Caucasians and Jews having the highest IQs and blacks the lowest (with African [Ethiopian and Soweto] black children's IQs ranging between 57 and 70). He used all kinds of spurious evidence to support his argument including the claim that there was a specific gene, the so called "Duffy blood group gene - a gene found in 43 percent of Caucasians and in only 1 per cent of Black Africans" (p. 323) to support his racial arguments.

It is unbelievable that there are those who still hold such views in this era of social progress. The implications of such views cannot be over-emphasized. It is a short leap from asserting that some people are intellectually superior to others based on racial differences to arguing that those from the supposedly "superior" races should be 'masters' and those from "inferior" races should be slaves and servants [Arendt (1998) calls them laborers (who prevalently use bodily brute strength) as opposed to workers (who use imagination and their minds to make lasting creations)].

We have seen the human consequences of such pernicious, racially motivated use of intellectual discourse to reach damaging conclusions in hundreds of years of slavery and direct partitioning and occupation of Africa, combined with reduction of the African people to a life of servitude during the colonial era, events that caused untold devastation of Africa and her people that continues to be felt to this day. No wonder Africans do not do so well in IQ tests. Any one who has been told that he/she is inferior over and over again for hundreds of years while at the same time being subjected to degradation and humiliation would end up believing it and doing as badly, even if we assume (and there is reason to doubt such an assumption) that intelligence tests created in the West are valid for the cultural/social contextual experiences of black populations in the world.

Combining Brand's argument of intellectual inferiority of some races with the individualistic attribution of poverty would lead to the implication that it is alright for those from the 'inferior' races to be poor because they lack the intelligence to succeed. This is probably the basis of the opposition to helping those who are poor. After all, the view of opponents to all kinds of welfare programs could be that the majority of the poor people come from supposedly 'inferior' races and therefore poverty for such individuals is the natural consequence of their station in life. It is a way of nature to wean them out of existence through natural selection, they may argue.

The seeming disdain for the poor has tended to spread throughout the world. For example, poor people in the third world suffered in the 1990s under the structural adjustment programs instituted by the World Bank (WB) and the International Monetary Fund (IMF) as a condition for third world countries to qualify for financial aid and loans from those institutions. Structural adjustment meant that governments were required to eliminate all or most of the social assistance programs such as public funding of education to make it affordable for the poor, free medical services, etc. In a world where the majority of the people live on less than a dollar a day, this was devastating.

The condition continues to be dire for the poor in those countries that are still dependent on those institutions for financial assistance.

It is therefore clear that how individuals perform in their occupational pursuit of political participation, whether as voting or elected citizens, has a significant impact on global issues such as material inequalities and poverty. We saw earlier that such occupational performance also affects what happens with regard to environmental protection. While doing research for this book, the views of a sample of individuals regarding the various global issues were assessed through a survey (Ikiugu, Anderson, & Anderson, 2007). Those views were correlated with the study participants' opinions regarding whether they were willing to change their occupational performance patterns and what they were willing to change in order to help resolve those issues. The study and findings are described in more detail at the end of the chapter.

I would like to emphasize here that as mentioned earlier, productivity is also related to other areas of occupational performance (leisure and self-maintenance). For example, materially poor people may have decreased opportunity to participate in certain types of occupations. Environmental destruction may also limit available leisure opportunities. Similarly, self maintenance may be affected by the economic status of an individual. Someone with limited funds may be unable to afford the means of mobility, such as a car, that may be needed in order to get around in the community in an endeavor to complete Instrumental Activities of Daily Living, such as shopping. The relationship between productivity and other categories of occupational performance will be discussed further in subsequent sections in this chapter.

Global Consequences of Individual Leisure Pursuits

Global issues such as poverty, diseases, material inequalities, and environmental destruction and pollution impact what individuals do for pleasure. Poverty may limit available choices for recreation. An example is my home location of Ruiri, in Meru, Kenya. Because of limited resources, the majority of people in the location cannot afford to go to the movies. There are no recreation centers, such as theme parks, where people can go with their families for relaxation. The environment is so limiting that the only available recreation is drinking cheap local beer and playing a local marble game known as *Bao*.

Ruiri is a good illustration of what Townsend and Wilcock (2004) refer to as occupational injustice which they conceive to be a lack of "enabling equal opportunity for meaningful and diverse occupations through redefining the way resources are shared...redistribution of money, access to environments that promote occupational development" and subsequent "lack of access to meaningful occupations" (p. 245). Occupational deprivation due to this injustice in turn has negative consequences on health. It contributes to stress occasioned by an individual's inability to exercise choice and participate in personally enjoyable occupations (Mernar, 2006). According to Mernar, there are biomarkers that can be used to measure this effect of occupational deprivation. These markers include physiological changes such as suppression of anabolism, decreased sensitivity to insulin (with increased risk of diabetes), constriction of blood vessels with subsequent hypertension etc., as well as psychosocial manifestations such as behavioral changes, decreased memory and learning ability, etc. Following is an explanation of how the above biomarkers result.

When an individual experiences occupational deprivation or imbalance (e.g. under-employment due to lack of opportunities or lack of resources necessary to engage in enjoyable occupations for productivity or leisure), occupational marginalization (inability to make choices about what, how, and when to do), and occupational alienation (lack of opportunities for engagement in desired occupations), stress results. The physiological response to stress whether caused by physical threat or occupational problems, is always "fight or flight." This involves activation of the Sympathetic Adrenomedullary system (SAM) and the Hypothalamic Pituitary Adrenocortical system (HPA).

When the SAM system is activated, nerve endings of the adrenal glands on the superior aspects of the kidneys are stimulated releasing catecholamine norepinephrine. Norepinephrine (and its derivative epinephrine) causes vaso-constriction, increased: release of glucose into the bloodstream (glycogenolysis); metabolism; heart-rate; dilatation of pupils of the eyes; etc., all in readiness for "fight or flight" reaction to the stress. At the same time, the HPA system is activated by stimulation of the hypothalamus of the brain to produce Corticotropin-Releasing Hormone (CRH) which stimulates the pituitary gland to release the Adrenocorticotropin Hormone (ACTH), eventually leading to production of cortisol.

Cortisol causes release of glucorcorticoids which act to suppress anabolism, decrease sensitivity to insulin with subsequent risk of diabetes, increased hypertension, etc., and also if their action is prolonged, impact the hypoccampal cells leading to increased rage, sexual drive, impaired memory and learning ability, impaired olfactory and gustatory functions, etc. In other words, prolonged hormonal imbalance that results from stress ["allostatic load" (Mernar, p. 210)] induced by occupation-related problems can result in both life-threatening physical diseases as well as significant cognitive and behavioral problems. In addition, some even suggest that lack of recreational opportunities in poor communities like Ruiri (discussed above) is responsible for the endemic incidence of sexually transmitted diseases, including AIDS since in the absence of anything else to do for leisure, sex becomes the only affordable form of recreation for the majority of the people.

As mentioned earlier, occupational performance in the area of productivity also influences issues such as poverty, material inequalities, and prevalence and consequences of diseases, since it can lead to decreased opportunities and occupational imbalance, deprivation, alienation, marginalization, etc. for some people. It can be argued that since those variables impact occupational performance in recreation, productivity is related to recreational occupations. Another good illustration of this relationship is that some recreational occupations are only accessible to individuals with money. Not many poor people can afford to play golf, participate in car racing, go sky-diving, etc. As a personal example, I love flying and would like to take flying lessons as a hobby. Apart from the fact that I may re-evaluate this desire in light of the environmental impact of flying (see earlier discussion of how airplanes contribute to global warming and climate change), flying lessons are so expensive that I cannot afford them right now. I have to focus on other priorities such as figuring how to pay for my children's college education. Thus, if I had been rich, I probably would have a pilot's license by now and flying frequently as a hobby. Therefore, there is a direct relationship between one's economic status (a result of productive occupational performance) and the type of recreational occupations that are open to the individual.

Besides the affordability of recreational occupations, as mentioned earlier, productive occupations have an impact on the environment. Destruction of the environment in turn influences recreational occupational performance. In a study involving a sample of adults living near Los Angeles, Eiswerth, Shaw, and Yen (2005) found that increased ozone levels have a statistically significant effect on the amount of time spent by individuals who are asthmatic on certain types of occupations (particularly outdoor chores and active leisure pursuits). Ozone problems are directly related to accumulation of hydro fluorocarbons in the atmosphere as a result of industrial activities (occupations in the productivity category).

Finally, there is a direct correlation between participation in leisure occupations and health (see discussion of the mechanism of this relationship above). Many studies indicate that engagement in physical leisure pursuits such as exercising and outdoor activities such as hiking is associated with better health and decreased incidences of chronic diseases and mortality rates (Bryan, Tremblay, Perez, et al., 2006; Ferrucci & Simonsick, 2006; James, 1992). Better health is in turn associated with increased productivity which as was mentioned earlier impacts the incidence of poverty, and material inequalities.

Therefore, recreational occupational performance is affected by consequences of human occupational performance in the area of productivity, such as material inequalities, environmental destruction, diseases, and poverty. Recreational occupational choices in turn affect the health of individuals and consequently their productivity. Besides, some of the recreational choices may negatively impact some of the global issues such as environmental destruction and pollution leading to global warming and climate change. One example is participation in the recreational occupation of car racing which contributes to global warming by facilitating burning of fossil fuels. Participation in the leisure occupation of golfing also means clearing large tracts of land in order to develop golf courses, which contributes to the process of desertification. Participation in the hobby of gardening on the other hand leads to cultivation of plants and trees which are good for the environment. In other words, an individual's choice and participation in leisure occupations impacts and is in turn impacted by the various global issues and also by performance in other occupational areas such as productivity.

Global Consequences of Self-Maintenance Occupations

There is a direct relationship between an individual's economic status and his/her ability to engage effectively in self-maintenance occupations. Consider the case of a person who works for minimum wages. The individual may be unable to save money to get him/her over hard economic times.

Suppose this person is involved in an accident in which he gets injured and is unable to work. He/she may have no medical insurance because he is unable to afford it on minimum wages. Therefore, he/she incurs debt in order to pay medical bills, which he/she cannot pay because of inability to work. The person is evicted from his/her domicile and becomes homeless. Homeless people lack resources such as clean, running water, facilities such as a shower, bath, etc. that are necessary for them to be able to take care of themselves effectively. They are most of the time pre-occupied with finding ways of staying alive with extremely limited resources (Rew, 2003).

It is therefore clear that homelessness, which to a large extent is caused by economic systems that create poverty and alienate the 'underclass' of poor people (Nelson, 2007), causes lack of resources necessary for effective self-care. Inability to take good care of oneself on the other hand is likely to cause deterioration of health leading to further inability to work, even if jobs are available, leading to decreased productivity, which exacerbates the problem of poverty even further. It has been observed that affordable housing is related to a family's capacity for economic mobility (Shlay, 1993) highlighting the impact of homelessness on occupational performance even further. If the family uses most of its resources to pay for housing, then it is not able to be economically self-sufficient. Members of a family who are rendered homeless do not only face challenges in their ability to take care of themselves (perform basic ADLs) but are also unable to engage in job-seeking and skill development occupations that would be necessary to make them more productive and economically self-sufficient.

As Shlay points out, any poverty reduction measures that aim at getting families off welfare assistance program without attempts to provide affordable housing is not a workable objective. Poverty impacts the ability to take care of oneself, and inability to take care of oneself may cause ill-health leading to decreased productivity, further poverty, and more challenges in ability to take care of oneself.

Global Consequences of Sexual Occupational Behavior

As mentioned earlier, sexual activity is not clearly identified in occupational therapy or occupational science literature as an occupation, although there often are allusions to the need for occupational therapists to pay attention to clients' sexual needs. Penna and Sheehy (2000) argued that sexual behavior is within the domain of occupational therapy since occupational therapists have to address clients' issues holistically. They conducted a study to find out if occupational therapists included sex education in their intervention programs with individuals with schizophrenia. They concluded from their study that many occupational therapists considered sex education to be within the domain of occupational therapy, but they were not providing it in their interventions.

Lowenwirth (2004) found that many individuals experienced difficulties with sexual expression post-stroke, and felt that their sexual issues were not addressed by any of the health care professionals who treated them. They indicated that they would appreciate an occupational therapist talking with them alone or with their spouses to address sexual issues arising out of their condition. Lloyd, William, and King (2005) described a program designed to provide health education to mentally ill individuals using an interactive didactic approach. However, others like Northcott and Chard (2000) found that many therapists, while believing that issues of sexuality should be dealt with during rehabilitation, were not sure whose responsibility it was to address those issues. Their findings implied that occupational therapists were not sure whether issues related to clients' sexual functioning fell within the domain of occupational therapy. One of the problems may be that occupational therapists are not adequately trained to handle clients' sexual issues. This would imply a need to incorporate such training in their educational curricular.

In this book, sexual pursuit is considered to be an important occupation that is very meaningful to human beings, and in whose engagement results consequences that impact global issues that are of concern to our planet and all life in it significantly. This occupation may be considered to be in the category of self-maintenance if we take the biological view of sex as a means of procreation and as a biological function like eating or relieving one-self, or it may be placed in the category of leisure if it is viewed as a fun occupation for human enjoyment and relaxation (Gecas & Libby, 1976). How sexual performance is interpreted depends on the individual person's perspective.

According to Gecas and Libby, the occupation of sexual intercourse has meaning to human beings beyond the biological aspects shared with other sexual animals because of the human capacity to attribute meaning to objects and the experiences associated with those objects. This capacity for symbolism provides sexual activity with two dimensions of meaning: the external, interpersonal (where the focus is on establishing a sense of trust and affection with another person), and the internal, intrapsychic dimension [referring to the motivational elements that produce arousal and commitment to the sexual activity] (Gagnon & Simon, 1973).

Thus, according to Gecas and Libby, there are four generally held perspectives about sex: 1) Traditional - This is the religious perspective in which sex outside marriage is considered to be sinful. In this view, sexual partners should have fidelity to each other. The purpose of sex is primarily procreation although there is an affective component to the activity. For those who hold this view, the meaning of sex is in its biological, procreative function, and therefore, for them, it may be classified in the self-maintenance category. 2) Romantic - In this perspective, sexual intercourse is considered acceptable between two people who are in love, whether or not they are married. In this view: "Love is a prerequisite to sexual relations" (p. 37). It is a means of strengthening the bond between two lovers, and when it is performed outside of love, it is reduced to bestiality.

The romantic perspective may also be seen as presenting sex as a self-maintenance occupation since being able to love and to be loved is one of the basic needs essential for human well-being (Maslow, 1970). 3) Recreational - In this perspective, sex is viewed as primarily a pleasurable activity and constraints of marriage and love as prerequisites for intercourse are de-emphasized. Rather, the objective of engaging in sexual activity is to experience pleasure and give pleasure to the partner. Sexual performance and the ability to give a sexual partner pleasure are valued to the exclusion of other factors. Viewed in this light, sex may be placed in the category of leisure occupations. 4) Utilitarian-Predatory - In this perspective, sex is a means to another end, such as making money, access to power, obtaining a position of prestige, etc. Sexual intercourse for individuals holding this perspective need not even be pleasurable. Viewed in this way, sexual activity may be classified in the category of productive occupations. Examples of individuals that hold this view would be sex workers (prostitutes). Their performance of sex is mostly a means of earning a living.

It is however important to point out that the above outlined perspectives are not mutually exclusive. One may hold the view of sex as acceptable in the context of love (romantic view), but also see sexual intercourse with a loved partner as a means of obtaining and giving pleasure, and thus a way of relaxing with the loved individual. The important point, however, is that one's view of sex may determine the outcome of his/her engagement in the activity. If one's primary perspective is that sex is a means of procreation (as is the common view of adherents to the Catholic faith); he/she is likely to engage in the activity in such a way that children result from the occupational pursuit. This may contribute to overpopulation.

The problem of overpopulation was discussed in chapter 2, where it was pointed out that it is likely to lead to abuse and destruction of the environment, as well as to negative economic consequences. Since population growth is related to depletion of resources such as farmland, water, and to pollution (Barr, Tropical Forest Trust, & McGrew, 2004; RAND, 2000), it follows that it is also related to decreased productivity since fewer available resources means that people have less means of producing. This in turn enhances the problems of material inequalities and poverty. Thus, engagement in the occupation of sex can lead to exacerbation of the problems of poverty, material inequalities, and environmental destruction.

This does not mean that people should not procreate. Reproduction is necessary if the human species is to continue to exist. However, the procreation should be planned such that we do not experience overpopulation to the extent that our planet is unable to support us. Indeed, we may already be having too many people on earth to the point where resources necessary to sustain us are being strained. Therefore, in pursuit of the enjoyable occupation of sex, a responsible person should be thinking about whether he/she really wants children to result from the activity. Also, people who become poor because of having too many children that they have trouble supporting have less ability to participate in political discourse, which affects governance (institutional functioning). Therefore, sexual occupation can be indirectly related to institutional failures.

On the other hand, those who hold the view that sexual activity is either a romantic or recreational occupation are more likely to take measures such as use of contraceptives in order to avoid conception in the process of enjoying sex, since their objective is not procreation but rather expression of love for a partner, or enjoyment and relaxation. Such an approach is therefore less likely to lead to over-population and all that that means, because such individuals have children only when they want them and know they can support them. The same argument can be made for those who approach sex from a utilitarian perspective. However, for such individuals, other negative consequences are likely to result from engagement in sexual activity. Those who see sex as a means of earning a living are likely to engage in sexual activity with multiple partners. Kane (1990) found in a study that some women engaged in sex with as many as 1800 men in the space of 6 months, and many times, they did not use protective measures such as condoms. Such behavior is likely to lead to the spread of sexually transmitted diseases such as AIDS, which themselves are global consequences of concern to humanity.

The conclusion that can be drawn from the above analysis is that sexual activity is an important human occupation, irrespective of the meaning it has to individuals engaging in it (whether it is viewed as a procreative, romantic, pleasurable, or utilitarian activity). Engagement in this occupation can have significant global consequences such as overpopulation, global material inequalities, poverty, environmental destruction, institutional functioning problems, etc. The specific consequences of engaging in the occupation depend on the individual's perspective regarding the meaning of sexual activity and action choices that one makes as a result of this view.

Global Consequences of Human Engagement in the Occupation of War

War is another human occupation that is not identified in occupational therapy or occupational science literature. Besides, it is a unique occupation because it does not fall under any of the categories of occupation as defined in occupational therapy or occupational science. It is not an occupation designed primarily for productivity. In fact, the whole purpose of war is killing and destruction. When one chooses to be a member of the military, militia, etc. that person knows that he/she may be called upon to kill somebody or bomb something. If sex, discussed above, is what Freud referred to as the *Eros* principle [the human instinct for love, tenderness, procreation, etc.] (Hergenhahn, 1997), war would be what he called *thanatos* or the instinct for destruction and self-annihilation. Therefore war, like sex, cannot be placed under the category of self-maintenance or leisure.

What is clear however is that war, probably more than any other human occupational pursuit, glaringly illustrates the global consequences of human choices and actions. This impact of human occupational choice will be illustrated with the most well known war going on in the world currently. In 2003, President George W. Bush declared war on Iraq, against opposition from much of the rest of the world. Knowledge of the history of the Arab nationhood and Iraq in particular would have made him and his strategists aware that this action was going to result in untold humanitarian catastrophe. For one, he would have realized that the experience of colonization in Iraq left a nation created artificially by the British colonizers with very tenable conglomeration of people from different religious sects (Shiites, Sunnis, and Kurds) with the Sunnis, again empowered right from the beginning by the British, lording it over the other sects who were seen as outsiders because they were not part of the former Ottoman empire (Romano, 2005).

Further, some knowledge of the history of Islam would have informed him and his strategists that the rift between the Shiites and Sunnis predates even Iraq as a nation state. The split goes back to the period after the death of Muhammad, the founder of Islam in the year 632, when the religion split into two factions: the followers of Ali, who was seen as belonging to Muhammad's lineage; and those of Abbasids, the Caliph whose dynasty built Baghdad (Tessler, 1994). The Ali partisans became known as the Shiites while Abbasids' followers became the Sunni Muslims. Knowledge of the above history would have given him some insight as to what to predict regarding the consequences of deposing Saddam, dictatorial as he was, by military force. Saddam kept the tenable, deeply divided nation together through iron rule. Indeed, Bush had been warned of the possible dire consequences. Jacques Chirac, French President, had warned of the fact that invasion of Iraq would destabilize the entire Middle East and cause terrorism to spread (Associated Press, 2007), which is exactly what happened.

Apparently, Bush and his strategists were not clear about this history or did not take the realities suggested by it into account when planning to invade Iraq. Either way, the US backed by a few other nations invaded Iraq and toppled Saddam. As feared by some, Iraq disintegrated into what can only be characterized as Civil war, mainly between Shiites and Sunnis, the historical rivals divided along religious sectarian lines. As a result, over 3000 US soldiers and thousands of Iraqis (24865 civilians between March 2003 and March 2005 according to the Iraq Body Count estimate, 98000 of civilians, soldiers, and insurgents according to the Lancet investigation estimates, 24000 between 2003 and 2004 according to the United Nations Development Program estimate, and 100000 according to a 2004 investigation) died (Johns Hopkins Bloomberg School of Public Health: Public Health News Center, 2004; Rai, 2005).

In addition, even though thousands of individuals, specifically the Shiites and the Kurds were displaced during the Saddam rule, the problem got even worse after the US-led war and occupation. According to the United Nations

High Commission for Refugees, since the beginning of the war in 2003, about 1.5 million Iraqis had fled Iraq for neighboring countries such as Jordan and Syria by the end of the year 2006 (Gillespie, 2006). Relegation to the status of a refugee meant losing any little wealth that one would have possessed, a means of livelihood, and abject poverty. In addition, destruction of infrastructure and resources meant that those who remained in the war-torn country lacked essential necessities. This was especially the case in a country like Iraq where the population depended very much on government rationing under the Oil for Food Program that was imposed on the country after the first Gulf War in 1992 (Anonymous, 2003).

This destruction exacerbates problems of child malnutrition among other humanitarian problems. The war in Iraq as is the case with most wars, caused thousands of deaths, destruction of infrastructure, loss of means of making a living, loss of homes, poverty, not to mention environmental problems. By making the decision to go to war in his occupational pursuit as the commander in Chief of the United States Armed Forces, George W. Bush had a far-reaching global impact that will continue to affect millions of people in the world for many years to come. Human engagement in the occupation of war is related to the global problems of destruction of human life and property, poverty, and environment.

Awareness of the Global Consequences of Occupational Performance by Individuals

As mentioned earlier, peoples' values, beliefs, opinions, and attitudes are likely to influence the way they make decisions and choices, and how they pursue daily occupations. For example, depending on what someone believes about the reason for existence of poverty [individual, structural, or fatalistic attribution] (Hunt, 2002; Shek, 2004), he/she she may feel compelled to do something to contribute towards reduction of the scourge in the world, or he/she may simply ignore the problem and assert that those who are poor deserve to be in that condition. This attitude would lead to a certain choice and pattern of engagement in occupations which in turn may have concrete and serious consequences for many people who are poor.

Since the thesis of this book is that individual choices and occupational pursuits have a significant impact on global issues that affect the planet and all life on it, it is important that we understand if peoples' beliefs are actually related to what they think they should do about the issues. To establish this relationship, a non-experimental type study with a survey design was conducted. The primary purpose of the study was to investigate whether there was "a relationship between attitudes held by individuals about select global problems (material inequalities, poverty, global warming/climate change, corruption and government failure, diseases, and overpopulation) and their willingness to change occupational performance patterns in order to impact those issues positively. We also sought to investigate the relationship between perceived human responsibility in causing those problems (here in after referred to simply as "human responsibility") and willingness to change occupational performance patterns" (Ikiugu, Anderson, & Anderson, 2007, p. 7).

Sixty one members of the American Occupational Therapy Association (AOTA) and the Society for the Study of Occupation: USA (SSO:USA), sampled from throughout the USA participated in the study. They responded to a questionnaire inquiring about the occupations in which they had engaged for two days; their opinions regarding the extent to which humans could be held accountable for causing various global problems; their judgment of the extent to which the occupations in which they had engaged for the two days had impacted the global problems; and their thoughts about how they could change their occupational performance patterns in order to influence those issues positively (see questionnaire used in the study in the appendix).

Data analysis revealed that respondents who had a negative individualistic poverty attribution (held the perception that people were poor because: they did not manage their money well; they wasted their money; they were lazy; it was their fate; or the state of poverty was their personal choice) were likely to be least willing to change their occupational performance patterns in order to influence global issues positively [z=-1.94, $p<.05$].

It was also found that there was a strong correlation between respondents' perception of human beings' responsibility in causing and influencing the state of selected global issues and their willingness to change their occupational performance patterns in order to influence those issues positively [$rs(N$=60)=.331, $p<.01$, 2-tailed]. Ordered probit models revealed that willingness to change occupational performance patterns in order to impact global issues positively was best explained by the level of education (z=2.03, $p<.05$), perceived human responsibility in causing the issues

(z=1.98, p<.055), and negative poverty attribution (see z and p values for this variable above).

Chi-square analysis indicated that out of the 61 respondents in the study, 31 (46%) were willing to change their occupational performance patterns in order to influence global issues positively. Twenty five respondents (37%) did not think they needed to change their occupational performance patterns and 5 (7%) were undecided. This distribution was statistically significant [X^2(2, N=61)=18.230, p<.01]. Further, Chi-square analysis indicated that a disproportionately large number of respondents (22/34 = 65% of those who responded to this item in the questionnaire) identified driving as an occupation whose performance needed to be changed [X^2(8, N=34)=99.941, p<.01]. Other occupations that were identified for change included watching less television in order to save electrical energy, being careful about doing laundry and taking shorter showers in order to save water, etc.

The above findings supported the hypothesis that there was a correlation between individuals' attitude towards global issues of concern and their willingness to change their occupational performance patterns in order to affect those issues positively. The findings also indicated that participants in this study largely recognized that human beings were responsible for causing and influencing the state of global issues discussed in this book. It was interesting to note that a positive (structuralist) poverty attribution also contributed towards readiness to change occupational performance patterns.

The findings suggested that if people were to be empowered to act proactively in order to contribute towards Gaia's healing through responsible choices and pursuit of daily occupations, there was need to educate them about the relationship between human occupational behavior and issues that could be considered to be both the symptoms and causes of Gaia's illness such as: poverty, material inequalities, diseases, corruption and government failure, environmental destruction and climate change, and population growth. Furthermore, the above findings provided evidence that recognition of the interconnectedness among these issues may motivate people to act (as indicated by willingness to change occupational performance due to positive attribution of poverty, which suggested that respondents recognized the connection between poverty and other issues of concern).

In addition, the finding that the level of education was significantly correlated to the willingness to change occupational performance was important because it suggested the need to advocate measures that would lead to an educated populace if we hope to impact the various global issues positively. Finally, the fact that a significant number of respondents recognized driving as an occupation that impacted Gaia negatively indicated that people in this sample were fairly aware of the global issues of concern and their etiology. For example, the finding implied that respondents recognized that adding to the accumulation of greenhouse gases through the occupation of driving contributed to environmental destruction and subsequent climate change. This increased awareness of the human role in making the global issues worse is an encouraging first step towards a possible cultural change that may lead to a positive change in the way individuals make choices and pursue daily occupations for the benefit of Gaia.

Of course the fact that this sample was derived exclusively from the community of occupational therapists and occupational scientists means that these findings cannot be generalized to the general population. It is possible that occupational therapists and occupational scientists are more attuned to the global issues of concern than the general public because they work largely with disadvantaged individuals who are most affected by these issues (such as the poor and those who are ill and may be more sensitive to environmental pollution). The study by Ikiugu et al. (2007) should therefore be replicated with other populations and with larger samples in order to verify the present results. Nevertheless, the findings are instructive in how an occupation-based program designed to help address the identified global issues could be structured. Such a program will be discussed in chapter 7. Interested readers may request a detailed report of the study findings discussed in this section from the author or any of the researchers who assisted in completing the study (see the complete reference). At the time of this writing, the paper reporting the above findings was under review for publication in a scholarly Journal.

Reflection Exercise #5

After reading chapter 5, articulate as honestly as possible your perception of how your occupational performance in the following areas affects either positively or negatively the global issues discussed in chapters 1 to 3:

1. Productivity (things you do to contribute to the economic well-being of yourself and the community, and the welfare of other people, including paid work, volunteering, home-making, care of a family member, etc.)

2. Self-maintenance including the type of food you eat, taking care of your health, personal hygiene, etc.

3. Leisure pursuits (things you do for pleasure)

4. Political participation

5. Sexual activity

6. How can you change the way you choose and perform any of the above occupations so as to influence the issues discussed in chapters one to three positively?

References

Agence France-Presse. (2007, February 25). Report: In US, record numbers are plunged into poverty. *USA Today*. Retrieved February 25, 2007, from http://www.usatoday.com/news/nation/2007-02-25-us-poverty_x.htm?csp=24.

American Occupational Therapy Association. (2002). Occupational therapy practice framework: Domain and process. *American Journal of Occupational Therapy*, 56, 609-639.

Anderson, S., Cavanagh, J., Hartman, C., Klinger, S., & Chan, S. (2004). *Executive excess 2004: Campaign contributions, outsourcing, unexpensed stock options and rising CEO pay - 11th annual CEO compensation survey.* Boston: Institute for Policy Studies and United for a Fair Economy.

Anelauskas, V. (2005). *Ending welfare, keeping poverty.* Retrieved January 3, 2007 from http:www.Pnews.org/PhPwiki/index.php/WelfarePoverty.

Anonymous. (2003). Rebuilding Iraq: Who's responsible? [Electronic version]. *Washington Report on Middle East Affairs, 22*(6). Retrieved January 20, 2007 from http://proquest.umi.com/pqdweb?did=3699826611&sid=2&Fmt=3&clientId=44616&RQT=309&VName=PQD.

Arendt, H. (1998). *The human condition.* Chicago: The University of Chicago Press. Original published in 1958 by the University of Chicago Press.

Ashford, S. (2001). Upward mobility, status inconsistency, and psychological health. *Journal of Social Psychology, 130*(1), 71-76.

Associated Press. (2007). Sunni group says Iraq government should be held responsible for any militia violence [Electronic version]. *USA Today*. Retrieved January 5, 2007 from http://www.usatoday.com/news/work/iraq/2007-01-05-iraq_x.htm?csp=24.

Australian Council of Super Investors, Inc. (ACSI) (2006). *CEO pay in the top 100 companies: 2005.* Melbourne, Australia: Author.

Baer, M., Bluestein, A., Chafkin, M., Gill, J., et al. (2006, November). The industrialist: You want to talk about skin in the game? It's time to meet Ray Anderson. *The eco-advantage Inc. Magazine*, 80-82.

Baum, C. M., & Christiansen, C. H. (2005). Person-environment-occupation-performance: An occupation-based framework for practice. In C.H. Christiansen & C.M. Baum (Eds), *Occupational therapy: Performance, participation, and well-being* (pp. 243-266). Thorofare, NJ: Slack.

Barr, R., Tropical Forest Trust, & McGrew, J. (2004). Landscape-level tree management in Meru Central District, Kenya. *Agroforestry in Landscape Mosaics Working Paper Series*. World Agroforestry Center, Tropical Resources Institute of Yale University, and the University of Georgia.

Brady, H. E., Verba, S., & Scholzman, K. L. (1995). Beyond SES: A resource model of political participation. *American Journal of Political Science Review, 89*(2), 271-294.

Brand, C. (1997). Race differences and race realism: Ten arguments for the existence of racial differences in intelligence and why we should welcome race realism. *The Mankind Quarterly, XXXVII*(3), 317-326.

Bryan, S. N., Tremblay, M. S., Perez, C. E., Arden, C. I., & Katzmarzyk, P. T. (2006). Physical activity and ethnicity: Evidence from the Canadian community health survey. *Canadian Journal of Public Health, 97*(4), 271-276.

Bush Administration. (n.d.). Working toward independence. No place, publisher not identified.

Capra, F. (1996). *A new scientific understanding of living systems: The web of life*. New York: Anchor Books/Doubleday.

Christiansen, C. (1994). Classification and study in occupation: A review and discussion of taxonomies. *Journal of Occupational Science, 1*(3), 3-20.

Cozzarelli, C., Wilkinson, A. V., & Tagler, M. J. (2001). Attitudes toward the poor and attributions for poverty. *Journal of Social Issues, 57*(2), 207-227.

Davis, R. (2006, December 27). Polar bears face meltdown [Electronic version]. *USA Today*. Retrieved December 28, 2006 from http://www.usatoday.com/news/washington/2006-12-27-polar-bears_x.htm?csp=24.

Eiswerth, M. E., Shaw, W. D., & Yen, S. T. (2005). Impacts of ozone on the activities of asthmatics: Revisiting the data [Electronic version]. *Journal of Environmental Management, 77*(1), 56. Retrieved January 20, 2007 from http://proquest.umi.com/pqdweb?did=870756371&sid=3&Fmt=2&clientId=44616&RQT=309&VName=PQD.

Ferrucci, L., & Simonsick, E. M. (2006). A little exercise. *The Journal of Gerontology, 61A*, 1154-1156.

Fuentes, R. (2005). *Human development report 2005 - Human development report office occasional paper: Poverty, pro-poor growth and simulated inequality reduction*. United Nations Development Program.

Gagnon, J. H., & Simon, W. (1973). *Sexual conduct*. Chicago: Aldine Publishing Co.

Gecas, V., & Libby, R. (1976). Sexual behavior as symbolic interaction. *Journal of Sex Research, 12*(1), 33-49.

Gillespie, K. (2006). Limbo land [Electronic version]. *The Jerusalem Report*. Retrieved January 20, 2007 from http://proquest.umi.com/pqweb?did-1170628371&sid=1&Fmt=3&clientId=44616&RQT=309&VName=PQD.

Githongo, J. (2005, November 22). *Report on my findings of graft in the government of Kenya*. Report Submitted to Mwai Kibaki, the President of the Republic of Kenya, State House, Nairobi.

Goransson. (2004, December 10). How not to bridge poverty gap. *Daily Nation Online*. Retrieved December 10, 2004 from http://www.nationmedia.com/dailynation/nmgcontententry.asp?category...

Gore, A. (Writer and Actor), West, B. (Actor), & Guggenheim, D. (Director). (2006). *An inconvenient truth* [Motion picture]. (Available from Paramount Classics, 2025 Broadway, Oakland)

Harrow, J. (1996). *Values and attitudes (in basics 2)*. Retrieved January 4, 2007 from http://members.aol.com/Jehanas/c_2val.html.

Heemskerk, M., Wilson, K., & Pavao-Zuckerman, M. (2003). Conceptual models as tools for communication across disciplines (Electronic version). *Conservation Ecology, 7*(3), 8. Retrieved December 22, 2006 from http://www.consecol.org/vol7/iss3/arts.

Hergenhahn, B. R. (1997). *An introduction to the history of psychology* (3rd ed.). Boston, MA: Brooks/Cole.

Hocking, C. (2000). Occupational science: A stock take of accumulated insights. *Journal of Occupational Science, 7*(2), 58-67.

Hunt, M. O. (2002). Religion, race/ethnicity, and beliefs about poverty. *Social Science Quarterly, 83*, 810-831.

Ikiugu, M. N. (2005). Meaningfulness of occupations as an occupational-life-trajectory attractor. *Journal of Occupational Science, 12*(2), 102-109.

Ikiugu, M. N., Anderson, L., & Anderson, W. (2007). *Occupational science in the service of GAIA: A study of the impact of human occupational behavior on global issues of our time*. Submitted for Publication.

Ikiugu, M. N., & Rosso, H. M. (2005). Understanding the occupational human being as a complex, dynamical, adaptive system. *Occupational Therapy in Health Care, 19*(4), 43-65.

James, O. (1992). Heading for the great outdoors: Leisure-time foodservice adapts to new times. *Restaurant business, 91*(14), 104.

Johns Hopkins Bloomberg School of Public Health: Public Health News Center. (2004). Iraq civilian death increase dramatically after invasion. Retrieved January 20, 2007 from http://www.jhsph.edu/PublicHealthNews/ Press_Releases/PR_2004/Burnham-iraq.html.

Kane, S. (1990). AIDS, addiction and condom use: Sources of sexual risk for heterosexual women. *The Journal of Sex Research, 27*(3), 427-444.

Kielhofner, G. (2002). Motives, patterns, and performance of occupation: Basic concepts. In G. Kielhofner (Ed.), *Model of Human Occupation* (13-27). Philadelphia: Lippincott, Williams, & Wilkins.

Kimenyi, M. S. (1987). Bureaucratic rents and political institutions. *Journal of Public Finance and Public Choice*, 39-49.

Knox, N. (2006, November 24). Wealth gap swallows up American dream. *USA Today*. Retrieved November 27, 2006 from http:// www.usatoday.com/money/perfi/housing/2006-11-24-luxary-homes-usat_x.htm?csp=26.

Law, M., Polatajko, H., Baptiste, S., & Townsend, E. (2002). Core concepts of occupational therapy. In E. Townsend (Ed), *Enabling occupation: An occupational therapy perspective* (pp. 29-56). Ottawa, ON: Canadian Association of Occupational Therapists.

Lloyd, C., Wiliams, P. L., & King, R. (2005). A sexual health program implemented in a psychiatric inpatient unit. *New Zealand Journal of Occupational Therapy, 52*(1), 26-32.

Lovelock, J. (2006). *The revenge of Gaia: Earth's climate crisis & the fate of humanity*. New York: Basic Books.

Lowenwirth, K. A. (2004). *Sexual dysfunction post stroke*. Ithaca, NY: Ithaca College.

Maslow, A. H. (1970). *Motivation and personality* (2nd ed.). New York: Harper & Row.

Mauro, P. (1998, March). Corruption: Causes, consequences, and agenda for further research. *Finance & Development*, 11-14.

Mbaku, J. M. (1996). Bureaucratic corruption in Africa: The futility of cleanups [Electronic version]. *Cato Journal, 16*(1), 1-15. Retrieved October 31, 2006 from http://www.cato.org/pubs/journal/cj16nl-6.html.

Mbaku, J. M. (1991). Property rights and rent seeking in South Africa. *Cato Journal, 11*(1), 135-150.

Mernar, T. J. (2006). Occupation, stress, and biomarkers: Measuring the impact of occupational injustice. *Journal of Occupational Science, 13*(3), 209-213.

Mosey, A. C. (1996). *Psychosocial components of occupational therapy.* New York: Lippincott, Williams, & Wilkins.

Mosey, A. C. (1970). *Three frames of reference for mental health.* Thorofare, NJ: Slack.

Namunane, B. (2006, November 7). Revealed: Graft costs Kenya Sh85bn per year. *Daily Nation Online.* Retrieved November 8, 2006 from http://www.nationmedia.com/dailynation/nmgcontententry.asp?category_id =1&newsid=85039.

Nelson, K. (2007). The child welfare response to youth violence and homelessness in the nineteenth century [Electronic version]. *Child Welfare, 74*(1), 56-70. Retrieved January 16, 2007 from the Psychology and Behavioral Sciences Collection database, http://web.ebscohost.com/ehost/delivery?vid=7&hid=17&sid=c14e02a2-9b63-4012-896f-...

Northcott, R, & Chard, G. (2000). Sexual aspects of rehabilitation: The client's perspective. *British Journal of Occupational Therapy, 63*(9), 412-418.

Odum, H. T. (1983). *Systems ecology: An introduction.* New York: John Wiley.

Penna, S., & Sheehy, K. (2000). Sex education and schizophrenia: Should occupational therapists offer sex education to people with schizophrenia? *Scandinavian Journal of Occupational Therapy, 7*(3), 126-131.

Personal Growth Center. (n.d.). The power of positive personal attitudes. Retrieved December 29, 2006 from http://www.gurusoftware.com/GuruNet/KnowledgeBase/Personal/AchievingLifeAttitudes.htm.

Rai, M. (2005). Iraq mortality. Retrieved January 20, 2007 from http://iraqmortality.org/iraq-mortality.

RAND. (2000). Population and environment: A complex relationship. *Population Matters: Policy Brief.* Retrieved November 20, 2006 from http://www.rand.org/pubs/research_briefs/RB5045/index1.html.

Rew, L. (2003). A theory of taking care of oneself grounded in experiences of homeless youth. *Nursing Research, 52*(4), 234-241.

Romano, D. (2005). Whose house is it anyway? IDP and refugee return in post-Saddam Iraq. *Journal of Refugee Studies, 18*(4), 430-453.

Sala-i-Martin, X. (2002). The disturbing "rise" of global income inequality. New York: Columbia University, Department of Economics Discussion Paper Series.

Satter, R. G. (2007, January 4). El Nino, greenhouse gases predicted to make 2007 hottest ever [Electronic version]. *USA Today.* Retrieved January

4, 2007 from http://www.usatoday.com/weather/climate/2007-01-04-climate-prediction_x.htm?csp=24.

Shek, D. T. (2004). Beliefs about the causes of poverty in parents and adolescents experiencing economic disadvantage in Hong Kong. *The Journal of Genetic Psychology, 165*(3), 272-291.

Shlay, A. B. (1993). Family self-sufficiency and housing. *Housing Policy Debate, 4*(3), 457-495.

Smith, A. (1776). An inquiry into the nature and causes of the wealth of nations [Electronic version]. Retrieved November 17, 2006 from http://socserv2.socsci.mcmaster.ca/~econ/ugcm/3113/smith/wealth/wea/bk01&bk.

Stoller, G. (2006, December 19). Concern grows over pollution from jets [Electronic version]. *USA Today*. Retrieved December 20, 2006 from http://www.usatoday.com/money/bitztravel/2006-12-18-jet-pollution-usat_x.htm?csp=24.

Tessler, M. (1994). *A history of the Israeli-Palestinian conflict.* Indianapolis, IN: Indiana University Press.

Townsend, E., & Wilcock, A. (2002). *Occupational justice.* In C.H. Christiansen & E.A.

Townsend (Eds,), *Introduction to occupation: The art and science of living* (pp. 243-273). Upper Saddle River, NJ: Prentice Hall.

Vasagar, J. (2005, February 24). Kenyan president faces rebellion on sleaze [Electronic version]. *The Guardian*. Retrieved January 4, 2007 from http://www.guardian.co.uk/print/0,,5133762-111242,00.html.

Verba, S., Schlozman, K. L., & Brady, H. E. (1995). *Voice and equality: Civic voluntarism in American politics.* Cambridge, MA: Harvard University Press.

Wilcock, A. (2006). *An occupational perspective of health* (2nd ed.). Thorofare, NJ: Slack.

PART III

PARTIAL PRESCRIPTION FOR GAIA'S HEALING: A PROPOSED CONCEPTUAL FRAMEWORK FOR THE SOLUTION OF GLOBAL PROBLEMS

As explained in the introduction, the rehabilitative approach used by occupational therapists is presented as an analogy for the problem-solving process proposed in this book to help solve Gaia's problems. When occupational therapists embark on a rehabilitation endeavor, they begin with information gathering in order to identify occupational performance issues of importance to the client, proceed to develop goals to address those issues, establish a plan of action to facilitate attainment of the goals, and implement the plan. The same logic is followed for the proposed occupation-based approach to resolution of global problems and the healing of Gaia proposed in the book. Therefore, in part I, extensive literature review was conducted in order to gather information and bring to light some of the pressing global issues that face humanity today. This phase was analogous to information gathering in clinical practice in an attempt to identify a client's pertinent occupational performance issues.

In part II, occupational science was introduced and an argument was made for the scientific discipline as a guide to action designed to solve issues identified in part I. This argument was based on the premise that to a large extent, the identified global issues facing humanity today are a result of the choices and performance of individuals as they pursue their daily occupations in the areas of productivity, self-maintenance, leisure, sex, and war. Human beings have the power to change the pertinent issues of concern by changing their occupational performance patterns as individuals. Also, it was pointed out that individuals are often not aware of the far-reaching implications of their occupational choices and the power they wield through such choices.

In part III and the final section of the book, a conceptual framework for the solution of the global issues will be presented. In chapter 6, it will be argued that the problem of our approach to resolution of prevailing global issues so far has been the tendency to emphasize "top-down" (also referred to as deductive) interventions without due focus on individual actions and building on them (also referred to as inductive approach). It will be pointed out that individual activity is the building block for systemic functioning and therefore individuals' actions can be harnessed to impact global problems. This framework is referred to as the "bottom-up" approach since the goal is to begin with individual actions to transform systems and impact global concerns (acting locally to obtain global effects). The two approaches (bottom-up and top-down) should therefore be used together in order to solve the global problems effectively.

In chapter 7, biographies of a few individuals who have impacted the world significantly through their actions will be analyzed to identify experiences, events, etc. that shaped their thought processes that led to their commitment to the service of the planet and humanity. Results of the analysis will offer guidelines about how ordinary citizens of the human race can be sensitized to commit to doing extraordinary things in order to influence the course of events in the world positively. The proposed framework for occupation-based solution of global issues and healing of Gaia will be two-pronged (consist of two approaches): 1) macro-level, in which suggestions for helpful systemic interventions will be made; and 2) micro-level (bottom-up), in which step by step directions of how individuals can teach themselves to choose and perform occupations reflectively in order to impact global issues positively will be presented.

Chapter 6
Occupation-Based Approach to Addressing Global Issues of Concern: A Proposed "Bottom-Up" Approach

The issues discussed in this book (poverty, material inequalities, destruction of the environment, diseases, institutional dysfunction, overpopulation, etc.) are not new. Humanity has been dealing with them for a long time. However, for some time now, the approaches that have been used to address these issues have been largely "top-down" (what I refer to as "deductive"). The assumption has been that the way to deal with them is to introduce system-based control mechanisms, such as policies to provide incentives or dis-incentives for certain actions. For example, it is assumed that if we want to increase innovation, we only need to provide tax-breaks (economic incentives) to individuals who engage in entrepreneurship.

The deductive approach to solving global problems may be traced back to the work of thinkers such as Adam Smith (1776) or probably even others before him. According to Smith, nations' wealth is built by human labor which produces the necessities needed by citizens. These necessities are exchanged with those from other nations. It is the division of labor among individuals based on their varying skills and talents that spurs national productivity and subsequent exchange of goods, the two key ingredients of commerce and economic prosperity. Furthermore, when more goods than are needed to meet the necessities of citizens are produced, accumulation of wealth within nations ensues.

In his thesis, Smith introduced one very basic proposition that revolutionized the approach to business and commerce. He postulated that

145

people did not work and produce out of the magnanimity of their hearts. Their motivation was not to help others. It was self-interest. People produced goods in order to accumulate personal wealth and to enjoy more of the luxuries of life. What people did in their productive pursuits was dependent on marketability of their goods (and/or services). Expenditure of labor and productivity were regulated by the supply and demand forces within the market system. As Smith put it:

> The market price of every particular commodity is regulated by the proportion between the *quantity which is actually brought to market* (supply), and the *demand* of those who are willing to pay the natural price of the commodity, or the whole value of the rent, labor, and profit, which must be paid in order to bring it thither. (p. 33, emphasis mine)

By introducing the twin constructs of *self-interest* as the motivator for human occupation related to productivity and the *supply and demand* dynamics of the *free market* as the invisible hand regulating how humans performed their occupations in an attempt to address that self-interest, it seemed that Smith eliminated human responsibility in making conscious choices and engaging in actions that produced benefits for society and the world beyond personal interest. In other words, he eliminated a broader view of individual responsibility to the world and the commonwealth, or self-transcendence if you will. Instead of encouraging individuals to act in accordance with what was best for humanity, the planet, and all living things, policy makers who saw Smith as their guide sought to introduce policies at systemic level, treating individuals within the system as if they were unthinking robots simply responding to behavioral mechanistic reward and punishment contingencies and nothing else. For example, to increase productivity of certain commodities, they proposed reducing taxes on productive endeavors related to production of those commodities to attract more people into that area of business.

It is assumed that individuals may not willingly feel the responsibility to make decisions to produce certain commodities just because they are good for humanity. In fact, those who demonstrate that responsibility by committing themselves to the service of humankind just because it is a good thing to do, such as teachers or nurses among many others, are the least rewarded by policy makers. In the ensuing system, generosity and desire to help others for the sake of helping have become signs of weakness rather than strengths to be admired. That is why individuals will not think twice about engaging in production of commodities that harm people or even cause death (e.g. cigarettes) rather than investing their time and talent on those things that in the long run increase the welfare of everybody on earth. As long as a commodity is in demand and makes money, and it is legal according to social consensus, many people are quite comfortable producing it.

Similarly, Karl Marx (1844), although diametrically opposed to Smith's capitalism, was nevertheless also deductive in his approach to human problems. In his view, workers mixed their labor with natural resources in order to produce commodities. However, in the capitalistic system, where labor was objectified (treated like a commodity that could be bought and sold), the worker was trapped in a vicious cycle in which the harder he or she worked, the more resources were appropriated for production, leaving less and less of such resources available for the worker to live on.

As an example, when more and more land was appropriated by the capitalist to produce food, which the worker's labor was actually responsible of producing, less land was available for the worker to cultivate and produce food for his/her consumption and to feed his/her family. The worker therefore became more and more dependent on the wages paid by the capitalist in order to sustain his/her life, for which he/she had to work increasingly harder. In this vicious cycle, the worker was increasingly dehumanized and alienated from his/her labor, the products of that labor, and other people.

In Marx's view, this de-humanization and alienation of the working class by the capitalistic class would result in class conflict which would eventually lead to overthrow of the capitalist system paving way for creation of a "class-less" society, which he contended was the most evolved form of society. In such a society, each person would be able to contribute to the best of his/her ability and get from society what was necessary to meet his/her needs. Thus, in Marx's view, as was the case in Smith's postulations, individual responsibility in choosing and acting was not emphasized. Rather, his concern was systemic social evolution through the dialectical process characterized by conflict between classes.

As argued in the beginning of the chapter, this top-down (deductive) approach to solution of global problems seems to be the adopted mode of addressing pertinent issues by many policy makers and even social scientists in current practice. This is best illustrated in the Bush administration's proposal about getting poor families off welfare and on their path to independence (Bush Administration, n.d.). In this policy, it was suggested that within 60 days, an individualized plan for a family on welfare be developed consisting of "constructive" activities designed to get the family into work and off welfare. The family's participation in the designed activities was to be monitored by the state and progress assessed on a regular basis. The mandated activities in which the family needed to participate had to include a minimum of 40 hours per week. The family also had to engage in at least 24 hours per week of unsubsidized employment. There were many other similar proposals in the administration's plan.

In all the above mentioned guidelines, the family's input regarding what it actually needed in order to engage in productive occupations and leave welfare seemed not to matter. There was no mention of any intention to work closely with the family to find out their circumstances, or what family members thought about how to get out of that situation. The attitude seemed to be that families were in welfare because they were somehow irresponsible and needed a strict prescription imposed by the state. They were also viewed as being in need of monitoring with threats of punishment (financial assistance being cut off) if they did not meet the mandated requirements in order to ensure that they were doing what the State deemed necessary in order for them to transition to work. Therefore, the plan consisted of systemic incentives that were supposed to coerce individual family members into acting according to the standards set by policy makers.

The same approach has been used in the third World countries as a means of encouraging economic responsibility. Structural adjustment programs pushed on the developing nations by the World Bank (WB) and the International Monetary Fund (IMF) in the 1980s and 1990s as a pre-condition for loans and financial assistance are still fresh in memory. The programs required governments to reduce spending on social programs such as education and health care, requiring what they referred to as "cost-sharing", where citizens were required to pay for a part of such services (introduction of what they termed, "user fees") irrespective of whether or not they could afford to do so.

Again, there was no consideration of the reality facing individual poor citizens that would have explained why they were so poor that they could not pay for services in the first place. There was no attempt to solicit their thoughts about what kind of assistance they needed in order to pull themselves out of poverty and pay for whatever services they needed. It was assumed that given a chance, individuals would prefer free services and therefore, they had to be forced to pay for what they needed. In an attempt to meet the conditions spelt out by the WB and IMF, many African and other third world countries implemented strategies that denied poor citizens access to essential services. In Ghana for example, "in the late 1980s, a visit to a specialist (medical) cost 10 times the daily wage. Unsurprisingly, there was 'substantial declines in the utilization of health care services'" (Anonymous, 2007, p. 15).

The same approach is being used to address current problems such as corruption. The Global Organization of Parliamentarians Against Corruption [GOPAC], a network of world parliamentarians committed to working together to eliminate corruption in their governments convened a seminar (Laurentian Seminar) in 1998, in collaboration with the World Bank Institute (WBI). The result of this seminar was a handbook (GOPAC, 2005) that detailed the parliamentarians' opinions about what caused corruption, how it affected governments, political participation, social structure, economics, etc., and how they thought it could be minimized or eliminated. Their thesis was that corruption was a consequence of weak government institutions, characterized by unbalanced governance ecology, where one branch of government (usually the executive) dominated the other branches (judiciary and legislature). The result of this imbalance was lack of accountability, transparency, and political participation. These circumstances were perceived to be the ones that bred corruption because without transparency and accountability, state officials could appropriate public resources with impunity and distribute them to supporters, cronies, etc. At the same time, opportunity for illegal payment by businesses and citizens who sought services and tariff reduction in order to do business and increase profit margins increased.

Based on the above premise, GOPAC proposed a multi-pronged approach to combating the vice, consisting of rebalancing the ecology of governance. This, they suggested, would be achieved by strengthening the judiciary, legislature, and the civil society so that those institutions provided the necessary oversight to the executive and the law was effectively legislated and enforced making it possible for corrupt practices to be punished as necessary. Therefore, their proposed strategies included legal reform, legislation of anti-corruption laws, establishment of a framework for parliamentary oversight on the executive, making the politicians accountable for decisions they made that hurt public interests, and establishing a strong civil society with a vibrant political participation where citizens effectively evaluated their governments and provided feedback accordingly. Once again, this plan did not seem to include helping individual citizens to think critically about their actions and how those actions impacted the practice of corruption positively or negatively.

In all the above discussed examples, the Newtonian type approach to issues (which is being referred to as top-down in this book) is used. This is the deductive logical problem solving approach based on the premise that if general principles are discovered that explain the cause and effect relationships among these issues, such principles can be applied in a system in order to effect change in individual cases. For example, in Adam Smith's theory, the law of supply and demand can be used to regulate production and pricing of commodities and services, and to distribute wealth in society. The individuals who are the actors in the system simply respond to the supply and demand dynamics to produce what is most needed in society and in the process build their own wealth and that of the nation. Whoever can respond to this supply and demand cause and effect relationship in the most efficacious way is rewarded by reaping more profits from his/her labor.

Similarly, in Karl Marx's dialectical view of social evolution, the general law of conflict between social classes is used to explain and predict social change. In the more contemporary examples discussed above, the Bush Administration's proposed program of facilitating transition of families from dependence on welfare to independence seemed to be based on the general proposition that poverty was caused by unwillingness of individuals to work and earn a living. Therefore, left alone, poor people would prefer free handouts rather than working for a living, hence, the need to condition public financial assistance on demonstrable efforts mandated by states to obtain and sustain employment.

Using the same deductive logic, GOPAC proposed that the general principle that explained the pernicious practice of corruption and other socio-political evils was what they referred to as: "The Unbalanced Ecology of Governance" (p. 8). By unbalanced ecology of governance they meant a government system in which the executive dominated and suppressed other institutions of governance such as the judiciary and the legislature. The imbalance in their view caused lack of transparency, accountability, and citizen political participation, a lack that was perceived to be a catalyst for corruption. Therefore, in their view, the solution was to re-balance governance by limiting executive power and strengthening other government institutions such as the judiciary and the legislature, as well as political participation, which in turn would increase oversight of governance practices leading to decreased corrupt practices.

In all the above cases, the individual is treated like an automaton, simply responding to purported laws of nature, whether such laws include supply and demand, conflict between forces of change and the status-quo, or unbalanced ecology of governance. It is true that general laws do apply in regulation of individual instances of behavior. It is undeniable that strong legislation against corruption with contemporaneous strengthening of the judiciary and the police force and other branches of the executive so that enacted laws are applied competently and fairly would make corruption unattractive. The benefits of the practice would not be worth the penalty when caught. This might make someone who is interested in engaging in corrupt practices think twice before succumbing to the temptation.

Education may be another systemically initiated approach that may be beneficial in helping individuals change their thinking patterns so that they make choices and perform occupations responsibly. It could be mandated that ethics be part of school curricula so that students are taught from their primary school years through college what it means to be responsible for: environmental preservation; the well-being of other people, other living things, and oneself; ethical and responsible citizenship in the world, etc.

If such education is started early when children's brains are still highly malleable, it may have a significant impact in changing the world's culture and social attitudes in the future so that individuals choose as a matter of fact to act in ways that impact the issues of concern positively more often. The above assertion is consistent with the findings by Ikiugu et al. (2007) that the higher the level of education, the more readily individuals were willing to change occupational performance patterns to influence global issues positively (see the complete report of findings in chapter 5). The effect of such an approach may be explainable from the perspective of what Eakman (2007) refers to as social complexity. In this view, by changing cultural attitudes through government education programs, occupational behavior patterns that are novel, consistent with planetary and human well-being, and transcend the level of any one individual, would emerge.

However, it is argued in this book that the deductive, systemic approach alone to solving global problems is incomplete. One may even argue that this approach, when used alone, presents a rather degrading view of the human being as irresponsible and as being no better than any other animal that simply responds to environmental conditions instinctively without much forethought. Such a view, in my opinion, is inaccurate and dehumanizing to those who are the referents of such thinking (such as poor people whose activities it is thought necessary to monitor in order to force them to take responsibility of working in order to earn a living). The assumption that significant changes only come out of systemic interventions based on general laws governing cause and effect relationships is not justified.

Even many of those system interventions originated from individuals. Smith was the author of the notion of "supply and demand" and "free market system" as a way of stimulating economic growth, ideas which changed human understanding of global economic activity for ever. Similarly, Karl Marx illuminated the idea of social evolution to a more just (classless) society. Others like Isaac Newton, Albert Einstein, and other scientists individually contributed to our current scientific view of the world. No legislation or official policy could have created Newton's *Principia Mathematica*, or Einstein's general theory of relativity. On the negative side, Adolph Hitler created Nazism which resulted in death of millions of people. Idi Amini of Uganda had the same effect in his dictatorial rule of Uganda. Each of the above mentioned individuals made choices and acted in the course of pursuing daily occupations, producing consequences (whether negative or positive) that affected humanity and the entire planet at a grand scale and for many years. As Einstein (1952, cited in Isaacson, 2007, p. 7) stated, "It is important to foster individuality...for only the individual can produce new ideas." This means that it is individuals who cause significant changes in our planet since such changes initially begin as individual ideas.

I prefer to agree with Socrates, the Greek philosopher whose view of humanity was more validating and granted individual humans more responsibility in impacting global issues. According to Socrates one needed to know virtue in order to be virtuous (The Radical Academy, 2007). To him, virtue was the way to happiness. Since all human beings wished to be happy, the only reason that one was unable to obtain happiness was because he/she did not know the way to this condition (which he postulated to be virtue as opposed to evil). In his view, no human being was inherently bad. Evil was in reality just ignorance. The way to eliminate evil was therefore by eliminating ignorance through facilitation of knowledge acquisition, since "knowledge is virtue, and improper conduct results from ignorance" (Hergenhahn, 1997, p. 36). If for example, one realized that eliminating poverty in the world is not only virtuous but leads to greater happiness because of reduced tendency towards conflicts, or that engaging in corruption could lead to institutional collapse that would eventually reduce one's ability to enjoy life, then one would be much more willing to act in such a way as to reduce those vices.

But what was this knowledge that was supposed to lead to virtue and happiness? According to Socrates, it involved knowledge of how to: manage one's estate efficiently (not for personal self-interest and pleasure as Smith would suggest, but to benefit friends so as to win their love); serve one's country in order to gain its honor and recognition; work the land in order to produce abundant crops; and train one's body so as to be physically efficient (Beck, 2007). It will be proposed in the next chapter that helping an individual to gain knowledge as postulated by Socrates through examination of the legacy that he/she wants to leave in the world and to act in accordance with that image (what I refer to as reflective occupational performance) may be one way of supplementing systemic approaches such as legislation in order to address global issues of concern to humanity more holistically.

When the Socratic component of virtue is added to Smith's idea of self-interest, which he thought was so crucial to building the wealth of nations, that self-interest is understood to include concern for the welfare of other humans, all living things, and our planet in general (deep ecology according to Naess). The way to happiness could be understood not in terms of material pleasures of Smith's version of self-interest, which Socrates thought were fleeting at best (Beck, 2007), but in terms of virtuous living focused on the well-being of other people, one's country, and the entire planet. A few examples will now be presented to demonstrate how reflective action by individuals can result in significant positive effects globally.

Acting Locally to Create Global Effects

Even today there are many examples of individuals who are quietly creating revolutions in the world, if only in a limited sense. For example, the case of Ray Anderson, Chairman of Interface (Baer, Bluestein, Chafkin, et al., 2006) was discussed in chapter 5. After Anderson realized how much his tile carpet making company was plundering natural resources and contributing to pollution of the environment, he made a conscious choice to change how the company did business. Through a number of measures designed to reduce wastage and use more environment-friendly technology, he was able to increase the profit margin for the company while reducing environmental destruction. But Anderson went beyond just making innovations for his company. He realized that he needed to do something to raise awareness to encourage other companies like his to do business differently. He attempted to do this by delivering talks all over the nation, sharing his experiences with others.

Anderson also demonstrated by example what it meant to live reflectively for the sake of the environment. For example, he chose to: drive a smaller car; "he now drives a Prius" (p. 81); and use alternative sources of energy that do not involve burning of fossil fuels or use of other more environmentally destructive sources of energy as demonstrated in his statement that: "We do have a weekend house in the mountains: It's off the grid, entirely solar powered" (p. 81). He also embarked on an environmental restoration program where: "Interface tries to offset the environmental cost of air travel with a program of tree-planting: some 62000 of them so far. (The rough calculus: a tree for every 2000 passenger air miles.)" (p. 81).

Through efforts of individuals such as Anderson, a revolution to embrace environment-friendly practices is already occurring in the business world. An example of this revolution is the efforts made by General Electric (GE) where according to Anderson: "Jeff Immelt has doubled GE's commitment to clean technologies-he's not doing it out of altruism alone. He's hearing his marketplace, just as we heard ours 12 years ago" (p. 81). By acting reflectively in pursuit of their daily occupations in the area of productivity, individuals such as Anderson and Immelt have impacted the global issue of environmental destruction and pollution positively, with far-reaching effects. Their examples illustrate the potential power of individuals at every level of participation in impacting global issues. Both Anderson and Immelt started revolutions in their companies as a response to the market call for change. This means that in pursuit of our occupation of consumption (shopping), we can cause significant change by demanding accountability in terms of the extent to which various companies address global issues of concern in the process of producing things that we consume.

Another individual who is trying to act to impact global issues through personal effort in pursuit of productive occupations is Yunus, the Bangladeshi economist who won the 2006 Nobel Peace Prize (Associated Press, 2006). Yunus decided to act on his belief that "we can create a poverty free world if we collectively believe in it" (p. 1 of 2). He also believed that eliminating poverty by putting more resources in fighting the scourge rather than on guns was the most effective way of fighting terrorism because poverty was, in his perception, the root of the evil. Acting on his beliefs, he established the Grameen Bank in 1983 specifically to provide micro-credit.

The bank aimed at lending small loans to poor Bangladeshi citizens who could not obtain loans from conventional banks because they did not qualify. The bank required no collateral in order to extend small loan facilities to these poor customers. Repayment was based on the honor system. Surprisingly (and contrary to common wisdom that would suggest that individuals would take advantage of the system), the repayment rate was 100%. Many poor Bangladeshis have since been able to establish small businesses and improve their lives significantly through loans from the bank. The idea spread throughout the world. This was an example of an individual who, by acting on his core beliefs while in pursuit of productive occupations, significantly impacted the global issue of poverty.

Scientific Rationale for Significant Global Effects of Individual Actions: Chaos/Complexity Theory as a Guide

The whole notion of GAIA proposed by Lovelock (1979, 2006) and Lovelock and Margulis (1974) is based on the notion that the earth and its atmosphere, upper layer of rocks, oceans, and the entire biosphere acts as a super-organism that regulates its internal environment, much like a live animal regulates its physiological functioning. This notion denotes an earth's ecology that acts as a complex, dynamical, adaptive system, interacting with its environment. This system consists of two parts: the thin spherical part extending from the interior of the earth to the upper atmosphere; and the interacting tissue of living organisms inhabiting the earth for billions of years (Tickell, 2006). Such a system can be understood by applying a conceptual framework derived from the chaos/complexity/dynamical systems theory.

Chaos/complexity/dynamical systems theory originated from the academic fields of mathematics, physics, meteorology, biology, chemistry, engineering, astronomy, and geography (Livneh & Parker, 2005). It is not a unified theory (Ikiugu, 2007; Ikiugu & Rosso, 2005; Livneh & Parker, 2005). However, there are common characteristics of phenomena that qualify them for definition as complex, adaptive, dynamical systems. Such characteristics include connectivity, diversity, non-linearity, self-organization, emergence, unpredictability, and fractality (Cooper, Spencer-Dawe, & Mclean, 2005;

Ikiugu, 2007; Ikiugu & Rosso, 2005; Livneh & Parker, 2005). Each of the above constructs will be briefly explained, and their implication to how we approach global issues will be discussed.

Connectivity refers to the property of dynamic relationships between *diverse* sub-systems within a system. Each agent within the system is connected to all the other agents in all their diversity in a dynamic co-evolution, through a mutual feedback system. This dynamic process is related to another property of complexity, known as *non-linearity*. Non-linearity refers to the fact that in complex systems, output is not equal to input, or as Linveh and Parker (2005, p. 20) put it, "cause and effect are not proportional, so that minor changes may result in large consequences..." This disproportion between input and output is a consequence of the connectivity property in complex systems. When new information is introduced (an event known as perturbation), it reverberates within the system (referred to as recursive effect) because of the dynamic feedback relationships between the agents in the system (Ikiugu, 2007; Mouck, 1998; Okes, 2003).

Through this feedback mechanism, the input is amplified, which accounts for disproportionately large output in comparison to the input. Weather is a good illustration of this non-linearity. Small variations in initial factors such as moisture in the atmosphere, direction and speed of the wind, and temperature can cause great differences in weather pattern outcomes (for example regarding whether precipitation will be in form of rainfall, a big storm such as a tornado or hurricane, a snow storm, etc.). This is because there are intricate complex relationships among wind, temperature, and moisture content that make any minor initial differences become amplified to produce significantly large differences in weather pattern outcomes. As will be pointed out later, this characteristic of complexity is crucial to our understanding of how our actions during occupational performance significantly impact the earth's entire ecosystem negatively or positively, with significant global consequences even when we think we are too small and our actions do not matter. It will also be pointed out how the characteristic may be central to explaining how we may act locally to impact global issues positively.

Related to connectivity and non-linearity are *unpredictability, self-organization*, and *emergence*. Following is a brief explanation of the relationship. First, because subsystems (agents) within complex dynamical systems are connected through mutual multiple feedback mechanisms (see figure 5-1 in chapter 5), and because of the non-linearity of these connections such that linear cause-and-effect relationships do not apply (hence disproportion between input and output such that small differences in initial factors or input can result in large differences in outcome or output), it is not possible to predict with certainty the outcome in such complex, dynamical systems. However, one can predict general outcome patterns over short periods of time. For example, local weather elements such as temperature, strength and direction of wind, moisture content, etc. can tell us whether we will get normal rainfall or a storm for may be up to five days in advance.

However, the specifics of such rainfall or storm, e.g. exactly how much the rainfall will be, or the exact path of the storm, may not be predictable even over such short time periods. The meteorologist can only predict that in general there is likelihood that there will be a fairly strong storm or heavy rainfall and give us the estimated range (such as between 2 and 6 inches of rainfall). This inability to predict the specific outcome in complex dynamical systems but the ability to foretell general patterns is very important in regard to our ability to predict the general outcomes of our actions in a complex dynamical, ecological system as will become apparent shortly.

The reason why complex dynamical systems have predictable patterns, even though the specifics are not discernible is because they have *trajectories* that are *self-similar* (Cambell, 1993; Ikiugu, 2007; Livneh & Parker, 2005). To understand the characteristic of self-similarity, it is important to grasp another construct, the *trajectory*. A trajectory is a path through which a dynamical system moves over time. For simple systems, such as a pendulum, a trajectory is an arc. The pendulum swings back and forth within the arc until it comes to rest at the point of equilibrium (known as systemic point-attractor). For complex systems, the trajectory is not so clear cut. Instead, it is characterized by variations over time. As an example, if we look at weather patterns over time, we see daily variations in temperature, direction and speed of wind, precipitation, etc. However, those patterns stay within certain constraints. In a certain area, temperatures may vary between 50 and 80 degrees Fahrenheit, never dropping below 50 degrees or rising above 80 degrees.

The variations discussed above form steady stable patterns that are predictable. We know that in winter, it will be cold, with temperatures ranging between negative 20 and positive 36 degrees. We may not know what the temperature will be on a specific day 2 months in advance, but we know it will probably be within that range. The variability range that constrains the system trajectory introduces the idea of *deterministic chaos* (Ikiugu, 2007; Mouck, 1998). The variables that form the trajectory seem chaotic on the surface, but underneath that seeming chaos, there is order characterized by clearly discernible recurring patterns of behavior. When examined on a small scale, these patterns resemble the over all trajectory of the system. In other words, the same pattern in the systemic structure is repeated over and over indefinitely (MacGill, 2006a). This characteristic is what is known as self-similarity, and a system presenting with this phenomenon is called a *fractal*.

Also, the self-similar patterns that emerge as a result of intense interaction between multiple agents and feedback loops between them are *emergent* phenomena that cannot usually be explained in terms of the qualities of the parts. They give the system characteristics that are unique and different from the specific qualities of any of the parts. Furthermore, the system moves within a space of operation known as *phase space*. The possibilities open to the system in this space determine its characteristics while the direction of its trajectory is determined by its *attractor*. The concept of attractor was introduced earlier. It is a point in the space of operation of the system at which the system will eventually end. As mentioned already, for simple systems such as a pendulum, the attractor is the point of equilibrium where the pendulum comes to rest.

For complex systems, the attractor may not be at a single point. For example, consider the solar system. The earth is orbiting around the sun, and the entire solar system including the sun and all its planets are orbiting in a trajectory within the galaxy. Thus, when the earth orbits around the sun, it does not arrive at the same point where it started because the whole system has already moved. In this case, where the object does not return to the same place, its attractor is referred to as a *strange attractor* (Lucas, 2004), which is a characteristic of complex systems. Furthermore, for such systems, there is usually more than one attractor.

Lucas gives the example of a putting green on a golf course. The attractor for the golf-ball is the golf hole. Once the ball gets at the edge of the hole, it falls inside. However, between the hole and the ball, there are a number of other attractors such as ridges that could deflect the ball towards another direction away from the hole. Therefore, the attractor that determines the trajectory of the system is ultimately the one to which the system is locked. Even this attractor is not a single point. Rather, it is like a basin consisting of multiple points. As an illustration, imagine a golf hole situated in a basin-like depression. When the ball gets at any point within this basin, it will drift into the hole. This depression is known as a *basin of attraction*, which is another characteristic of chaotic systems (Cambell, 1993; Ikiugu, 2007; Ikiugu & Rosso, 2005).

The construct of multiple attractors for a complex system introduces the idea of causation. In other words, what causes the system to lock on to a certain attractor or a particular basin of attraction? The answer is that internal or external changes in systemic variables determine which attractor becomes the strongest. In our golf example, the golfer hits the ball towards the hole, making the hole the attractor to which the ball system locks. In this case, external force is introduced causing the system to lock on to a certain attractor. Therefore, external or internal forces determine the direction of a system, pushing it towards a certain attractor.

However, sometimes, the force may not be adequate to push the system all the way to the end. Imagine a golfer who hits the ball but not strongly enough to get it all the way to the hole. The ball stops somewhere in the middle and he/she has to hit it again, and again, and so forth. At every point where the ball stops, there may be a ridge which may force it to go one way or another. At these points, it may continue to coast towards the hole, or it may drift in a completely different direction. In this analogy, two principles seem apparent (MacGill, 2006b): 1) Small changes build up to the tipping point which when reached (e.g. when the ball reaches the edge of the hole), the system accelerates in a certain direction; and 2) There are critical points in the path of the system's trajectory which force the system to make a choice between one direction or another. These critical points are referred to as *bifurcation points*, and are analogically represented by a fork on a road (Livneh & Parker, 2005; Mouck, 1998). The forces that push the system towards a certain direction, whether external or internal, are referred to as *Perturbations* (Ikiugu, 2007; Oakes, 2003).

When the perturbations referred to above push the system away from the point of equilibrium (indicated by stable steady patterns) the system is resistant to change. However, as the system moves further and further away from equilibrium, a critical point is eventually reached where it is pushed over the edge. At this point, there are two possible outcomes: 1) chaos ensues where the system spirals out of control; or 2) the system reorganizes itself at a higher level of functioning, making it possible to respond adaptively to environmental changes (perturbations) that have caused it to move out of its *stable steady state*. In other words, at the point of a major bifurcation, "Relationships 'shift and change, often as a result of *self-organization*'" leading to *emergence* of complex structures to enable the system *"to cope with or manipulate"* (Gatrell, 2003, p. 6) its environment (emphasis mine).

Complex, non-linear open systems "respond to perturbations by organizing into emergent form which cannot be predicted from knowledge only of the system parts" (p. 8). Consistent with the characteristic of sensitivity to initial conditions discussed earlier, small differences in combination of perturbation factors that lead to self-organization to a new level of systemic functioning can cause large differences in outcome, because both the 'errors' and 'functional' variables are amplified in the system as explained earlier. This possibility of rapid change and self-organization to a new functional state when a system reaches a tipping point, combined with sensitivity to combination of perturbation factors makes it imperative that the functioning of individual agents within the system be considered to be critical. Change in functioning of any one agent within the system can push the entire system to that critical point where it either spirals out of control or self-organizes to a new functional level (whether adaptive or maladaptive).

The above is a brief explication of the chaos/complexity perspective. It is by no means a detailed discourse on the theory. Readers can refer to the following sources for a more detailed discussion: Bassingthwaighte, Liebovich, and West (1994), Cambell (1993), Cooper, Spencer-Dawe, and Mclean (2005), Gatrell (2003), Livneh and Parker (2005), Lucas (2006, 2004), and MacGill (2006a, 2006b, 2006c, 2006d) among others. Readers may also find the following internet web-link useful: http://complexity.orcon.net.nz/html.

Implications of the Complexity/Chaos Theory Perspective to Global Issues of Concern

The earth and its ecosystem, or what Lovelock calls Gaia, can be understood to be a complex, open, dynamical, adaptive system. The agents within the system, including the rocky crust, living tissue (which includes humans and all other biological species), and the atmosphere are in a state of intense interaction with each other with feedback loops between them. This intense interaction results in emergence of planetary characteristics such as the weather, as well as intra-planetary phenomena discussed earlier in this book such as poverty, wealth inequalities, environmental destruction and global warming, institutional failures, population problems, etc. In other words, these global issues may be seen as emerging from the intense interaction between all the human and non-human agents within the planetary system and cannot be explained by understanding the characteristics of any one agent in the system. They transcend individual characteristics of any of the agents because they emerge "at the group level" and "cannot be understood at the level of any one individual (agent)" (Eakman, 2007, p. 87). However, each individual's behavioral patterns are important because they influence the emergent characteristics of the system. This is because they "mutually affect each other, and in doing so generate novel behavior for the system as a whole" (p. 87).

Furthermore, the system may be understood, for the most part, to be in a stable, steady, state. It is characterized by patterns, whose specificity cannot be predicted with certainty, but give us stable generalities. For example, we know in general terms that we should expect cold winters and warm summers, rainy and dry seasons, etc. However, human activities, including destruction of forests to create farming land and housing facilities, and commercial activities that are fuelled by burning fossil fuels and use of hydro-carbon-based raw materials are pushing the system further and further away from equilibrium (a stable, steady, state). Given these circumstances, it can be expected that very soon, the system will reach a critical point (it is probably there already) where it will either spiral out of control or re-organize itself into a new functional state where new characteristics will emerge.

As mentioned earlier, this self-organization may not be a problem for the planet itself. It would involve a shift in relationships between the agents. The emerging characteristics of the new steady stable state may not support a "fitness landscape" (Lucas, 2004) consistent with human survival. The human cell-chemistry may not be well equipped for extreme variations in temperature or the gas composition that may emerge in the new system. The result might be extinction of human species although some other biological life-forms that are more adapted to the new system's characteristics may emerge and prosper.

The other factor to consider is that as a system, Gaia is sensitive to initial conditions. Small differences in how variables within the system combine can have large differences in outcome. This means that actions of agents within the system, however minor, can have significant implications regarding the trajectory that our planetary ecosystem takes. This sensitivity to initial conditions is what gives immense power to individuals within the system. It means that small changes in how individuals choose to pursue their daily occupations can have large differences in whether the entire system is pushed to a critical point and subsequently whether it goes into a new state of self-organization which may lead to emergence of characteristics not conducive to human survival.

Also, although we cannot predict the specific systemic outcome of a certain trajectory, we can predict the result in general terms. Certain conditions would indicate to a keen observer that certain actions may lead to the system spiraling out of control. A real life example of this spiraling is evident in the Iraq situation. I listened to a speaker on CSPAN television who was explaining how the US conservative administration thought that their intervention in Iraq would lead to a democracy in the Middle-East that would spur the rest of the region to transform into a system that would be democratic, stable, and advantageous to the West. As he explained, the reasoning was that although Saddam Hussein was a ruthless dictator, he had managed to create a secular Arab Nation, with a thriving middle class, in which women were relatively liberated, and Islamic theocracy was largely absent in public life. Therefore, they reasoned, all that was needed was to replace the dictator with a democratically elected government and Iraq would be a democracy that would be a replica of the USA.

This thinking was largely linear (Newtonian) and simplistic. The authors of the strategy forgot to take into account the deep historical rivalries between Shiism and Sunnism, the Israeli-Palestinian problem, and the neighboring states' affiliations and influence on factions within Iraq, factors that made prediction of outcome of the intervention impossible. If they had been cognizant of the complexity of the situation, their approach to the situation would have been more cautious. Someone who had a respect for complexity would have predicted that if the dictatorship that kept the different factions in line was replaced from outside without internal agents being ready for the change, and without addressing other issues of contention such as the Israeli-Palestinian conflict, the entire system was likely to spiral out of control.

In essence, that is what happened. Once the strict dictatorship was removed, what resulted was not a functional democracy but rather chaos. Compare that outcome with the Kenyan transition. Kenya consists of about 42 ethnicities. Similar to Iraq, Kenyan people were subjected to a dictatorship for decades. However, because the democratic drive did not originate from outside but rather from inside the system, whereby factions within had decided that they needed to work together in order to obtain a common objective, democratization did not result in civil war and chaos. Instead, the transition to democracy was smooth and successful.

In terms of more global issues, not many people can argue with the fact that if we continue in the current trajectory where humans are destroying the landscape and the environment, eventually a point will be reached where the system will spiral out of control. Temperature patterns and atmospheric composition will be altered such that biological life as we know it today will probably not be viable. On the other hand, if humans change their actions immediately and begin rebuilding the earth's surface by planting trees and letting the earth replenish and heal itself; and if they stop polluting the environment; there is a chance that even if the planetary ecosystem goes into a new state of self-organization, it will continue to support life, including humans. In other words, actions on the part of each of us humans, no matter how small, can cause that small difference in combination of variables that will determine whether the result will be large differences in positive or negative outcomes in the future.

The conclusion reached above means that in addition to deductive efforts to influence issues positively through policy making and legislation (e.g. providing more strict guidelines to regulate industrial emissions), equal attention should be paid to education of individuals regarding their responsibility in shaping the future of this planet which is our collective home. Individuals should be educated to live and act reflectively with the hope that this will translate into responsible choices in pursuit of daily occupations. In this regard, we should go even further than Lovelock's (2006) proposal that action begin at the national level rather than waiting for world consensus. We should assert that action be encouraged at the individual level, consistent with Edward's proposal that the key to ending poverty, disease, global warming, and other pressing issues of our time is grassroots action (Lawrence, 2006). Individual action is also one of the strategies proposed by Gore (2006) in his award winning documentary[v].

Grassroots action should begin with helping individuals realize that how they make decisions and act as they pursue their daily occupations has consequences that go beyond their individual lives. This education is crucial considering the importance of education in influencing readiness to change as was found in our study (Ikiugu et al., 2007), and the complacency of many people as revealed in other studies where it was found that many people are not politically engaged even though politics affects every aspect of their lives significantly (Verba, Schlozman, & Brady, 1995). This awareness would empower people so that they are able to act consciously to alter the circumstances of our planet and all life in it for the better.

By empowerment is meant giving individuals: intellectual tools (such as critical thinking skills) necessary to enable them to make decisions; access to information and resources so that they are in a position to make meaningful choices consistent with their interest in a comfortable existence in this planet; ability to act assertively so that they are able to make a difference through individual actions; ability to redefine themselves, their abilities, and their relationship to power and authority; and ability to impact their lives and their communities (Chamberlin, 1997). That is the overall thesis of this book.

Reflection Exercise #6

After reading chapter 6, complete the following exercise:

A friend tries to ridicule your efforts by telling you that your actions cannot affect world issues significantly because you are a single individual, who is not famous, and whose job is not high profile or influential in the world. Using information obtained from chapter six explain to your friend why actions of individuals like yourself in pursuit of their daily occupations can significantly change the course of events in the world.

References

Anonymous. (2007). How IMF, World Bank failed Africa. *New African, 458*, 12-16.

Associated Press. (2006). Yunus claims Nobel Peace prize. *USA Today*. Retrieved December 12, 2006, from http://www.usatoday.com/news/world/2006-12-10-nobel-prizes_x.htm?csp=24.

Baer, M., Bluestein, A., Chafkin, M., Gill, J., et al. (2006). The green 50 - The industrialist: You want to talk about skin in the game? It's time to meet Ray Anderson. *The Eco-Advantage Inc. Magazine, 28*(11), 80-81.

Bassingthwaighte, J. B., Liebovich, L. S., & West, B. J. (1994). Fractal physiology. New York: 1994.

Beck, S. (2007). Greece & Rome to 30 BC: Ethics of civilization (vol. 4). Chapter retrieved from *About*, February 2, 2007, from http://ancienthistory.about.com/gi/dynamics/offsite.htm?.zi=1/XJ/Ya&sdn=ancienthistory&sdn=education&tm=24&gps=68_6_1276_605&f=00&tt=14&bt=1&bts=1&zu=http%3A//www.San.beck.org/EC21.Socrates.html.

Bush Administration. (n.d.). *Working toward independence*. No Place.

Cambel, A. B. (1993). *Applied chaos theory*. New York: Academic Press.

Chamberlin, J. (1997). A working definition of empowerment. *Psychiatric Rehabilitation Journal, 20*(4), 43-46.

Cooper, H., Spencer-Dawe, E., & McLean, E. (2005). Beginning the process of teamwork: Design, implementation and evaluation of an inter-professional education intervention for first year undergraduate students. *Journal of Interprofessional Care, 19*(5), 492-508.

Eakman, A. (2007). Occupation and social complexity. *Journal of Occupational Science, 14*(2), 82-91.

Gatrell, A. C. (2003). *Complexity theory and geographies of health: A modern and global synthesis?* Unpublished manuscript, Institute for Health Research, Lancaster University, Lancaster, Great Britain.

Global Organization of Parliamentarians Against Corruption. (2005). *Controlling corruption: A parliamentarian's handbook*. No place: World Bank Institute (WBI).

Gore, A. (Writer and Actor), West, B. (Actor), & Guggenheim, D. (Director). (2006). *An inconvenient truth* [Motion picture]. (Available from Paramount Classics, 2025 Broadway, Oakland)

Hergenhahn, B. R. (1997). *An introduction to the history of psychology* (3rd ed.). New York: Brooks/Cole.

Ikiugu, M. N. (2007). *Psychosocial conceptual practice models in occupational therapy: Building adaptive capability*. St. Louis, MO: Mosby.

Ikiugu, M. N., Anderson, L., & Anderson, W. (2007). [Occupational science in the service of GAIA: A study of the impact of human occupational behavior on global issues of our time]. Unpublished raw data.

Ikiugu, M. N., & Rosso, H. M. (2005). Understanding the occupational human being as a complex, dynamical, adaptive system. *Occupational Therapy in Health Care, 19*(4), 43-65.

Isaacson, W. (2007). *Einstein: His life and universe*. New York: Simon & Schuster.

Lawrence, J. (2006). Edwards launches presidential bid from New Orleans. *USA Today*. Retrieved December 28, 2006, from http://www.usatoday.com/news/washington/2006-12-28-edwards-2008_x.htm?csp=24.

Live Earth. (2007). *I pledge*: Retrieved July 7, 2007, from http://www.liveearth.org - SaveOurSelves-M...

Livneh, H., & Parker, R. M. (2005). Psychological adaptation to disability: Perspectives from chaos and complexity theory. *Rehabilitation Counseling Bulletin, 49*(1), 17-28.

Lovelock, J. (2006). *The revenge of the Gaia: Earth's climate crisis & the fate of humanity*. New York: Basic Books.

Lovelock, J. (1979). *Gaia: A new look at life on earth*. Oxford: Oxford University Press.

Lovelock, J., & Margulis, L. (1974). Atmospheric homeostasis by and for the biosphere: The gaia hypothesis. *Tellus, 26*, 2-9.

Lucas, C. (2006). *Emergence and evolution constraints on form*. Retrieved February 10, 2007 from http://www.caresco.org/emerge.htm.

Lucas, C. (2004). *Attractors everywhere - Order from chaos*. Retrieved February 10, 2007, from http:www.calresco.org/attractor.htm.

MacGill, V. (2006a). *Complexity pages: Exploring the new science of chaos and complexity - Introduction*. Retrieved February 10, 2007, from http://complexity.orcon.net.nz/intro.html.

MacGill, V. (2006b). *Complexity pages: Exploring the new science of chaos and complexity - History*. Retrieved February 10, 2007, from http://complexity.orcon.net.nz/histoty.html.

MacGill, V. (2006c). *Complexity pages: Exploring the new science of chaos and complexity - Fractals.* Retrieved February 10, 2007, from http://complexity.orcon.net.nz/fractals.html.

MacGill, V. (2006d). *Complexity pages: Exploring the new science of chaos and complexity - Mandelbrot set.* Retrieved February 10, 2007, from http://complexity.orcon.net.nz/mandelbrotset.html.

Marx, K. (1844). *Economic and philosophical manuscripts: First manuscript - Wages of labor* [Electronic version]. Retrieved July 2005, from http://eserver.org:16080/marx/1944-ep.manuscripts/1st.manuscript/1-labor.wages.txt.

Mouck, T. (1998). Capital markets research and real world complexity: The emerging challenge of chaos theory. *Accounting Organizations and Society, 23*(2), 189-215.

Okes, D. (2003). Complexity theory simplifies choices. *Quality Progress, 36*(7), 35-37.

Ollison, R. D. (2007, July 7). Climate in D.C. changes: Gore stages his Live Earth concert [Electronic version]. *Baltimore Sun.* Retrieved July 7, 2007, from http://www.baltimoresun.com/features/lifestyle/bal-to.live07,0,5435164.Story?coll=bal-artislife-today.

Radical Academy. (2007). *Classical philosophers - The great thinkers of Western philosophy: The philosophy of Socrates.* Retrieved February 2, 2007, from http://www.radicalacademy.com/philsocrates.htm.

Smith, A. (1776). *An inquiry into the nature and causes of the wealth of nations* [Electronic version]. Retrieved November 17, 2006, from http://socserv2.socsci.mcmaster.ca/~econ/ugcm/3113/smith/wealth/wea/bk01&bk04. Original published in Oxford by Clarendon Press.

Tickell, C. (2006). Foreword. In J. Lovelock, *The revenge of Gaia: Earth's climate crisis & the fate of humanity* (pp. xv-xvii). New York: Basic Books.

Verba, S., Schlozman, K. L., & Brady, H. E. (1995). *Voice and equality: Civic volunteerism in American politics.* Cambridge, MA: Harvard University Press.

Chapter 7
Occupation-Based Approach to Global Issues: A Suggested Conceptual Framework

It was pointed out in chapter 6 that since the planet and its ecology may be considered to be a complex, dynamical, adaptive system, individual choices and actions in pursuit of daily occupations in productivity, self-maintenance, leisure, sexual activity, and war have a far reaching impact on global issues (poverty, material inequalities, environmental destruction and global warming/climate change, overpopulation, and institutional failures). This is the case because of the connectivity and intricate feedback loops between humans and ecological agents such as other living organisms, geological and weather systems, etc. Therefore, it was argued that if we truly want to resolve the global issues of concern, we must emphasize "reflective living" where individuals are encouraged to choose and act in such a way that their occupational performance is consistent with improvement of the issues in question.

In 1961, an occupational therapist by the name of Mary Reilly, who is considered to be one of the most influential leaders in the profession, made a bold statement that: "One of the greatest ideas in 20th century medicine is that man, by use of his own hands, as they are directed by the mind and energized by the will, can affect the state of his own health" (Reilly, 1962, p. 2). Based on Reilly's assertion, the arguments made in earlier chapters will be expounded further in this chapter with a proposition that occupational performance is a powerful approach to resolution of the various global issues of concern to humanity and our planet. I would like to restate Mary Reilly's proposition as follows: One of the greatest ideas in the 21st century approach to global issues confronting humankind and the planet earth is that humans, by the use of their own hands as they perform daily occupations, directed by the mind and energized by the will, can affect the state of health of not only humans but of all living things and the planet's ecological system at large.

The reader should note in the above proposition that the power of human hands in performance of occupations and shaping of the environment can only be realized under the direction of the mind and facilitation by the will. The mind is crucial for instrumental action (problem solving and creativity) and the will is essential for arousal of passion necessary for meaningful human action. This connection between the mind and occupational performance is consistent with the notion of belief as a rule for action in the philosophy of pragmatism (Peirce, 1955), since belief is a product of the mind that is a necessary ingredient for that passion that energizes the activity of our hands. In trying to develop a way of addressing global issues through occupational performance, we can derive constructs from the philosophy of pragmatism, and especially from the idea advanced by John Dewey that the mind is an instrument that humans use as a tool to assist them shape the environment and make it their home (Dewey, 1960, 1981; Kennedy, 1996). To borrow from Dewey, in our conceptual framework:

> Intelligence becomes a 'fact' open to observation like any other natural fact. It is a distinctive kind of behavior, a certain mode of interaction between the live creature and its environment. That mode is one in which the organism in its present takes account of future consequences. (Kennedy, 1996, p. 333)

Similar to the pragmatists, and particularly John Dewey, we can view the role of the mind in human interaction with the environment as that of directing one's activities based on beliefs about self, other people, the environment, and the entire world. This view is especially clear in Dewey's notion that:

> The method of intelligence is the pragmatic or instrumentalist method. This method of persistently testing the meaning and worth of ideas, customs, institutions, in the light of their consequences, not just the immediate personal consequences, but their broad social consequences, leads us to the conception of a new society...(p. 334)

This pragmatic view is consistent with the assertion in chapter 5 that occupational choices and performance patterns are based on a person's attitudes, opinions, values, and beliefs. Attitudes, opinions, and values constitute "beliefs" in the pragmatic sense, because they become the "rules for action" as proposed by Peirce (1955).

The above perspective provides the rationale for the proposition that one of the best ways to resolve pertinent global issues may be by individual self-education about how to explore one's cognitions (particularly beliefs about self, other people, society, and the planet), choices and occupational performance patterns originating from those cognitions, and the global consequences of that performance. However, before embarking on that endeavor, it may be helpful to reflect a little on what would constitute that self-education. To achieve that objective, I completed a small investigation exploring some of the characteristics of individuals who have had significant influence on events in the world. The results of the investigation will be discussed next.

Ordinary People Who Do Extraordinary Things: A Brief Exploratory Study of the Characteristics of Individuals Who Have Changed the World

Study Methods

In this small exploratory study, naturalistic type methodology with phenomenological and heuristic designs (Creswell, 1998; Depoy & Gitlin, 2005; Giorgi, 1985; Moustakas, 1990; Speziale & Carpenter, 2003) was used. Using the phenomenological design, personal experiences, thoughts, beliefs, values, and opinions of the individuals constituting the study sample as gleaned from their written speeches were analyzed. The heuristic design was used to interpret written biographies with a view to identifying experiences and backgrounds that informed the study participants' perceptions, opinions, beliefs, values, and actions. This methodology was "autobiographic" (Kahakalau, 2004, p. 22) and therefore was most suitable for this study. As is typical in heuristic research (Moustakas, 1981), I used my own experiences as a person motivated to contribute to the solution of problems facing humanity to interpret the biographies.

Study Participants

Before completing the study, I requested colleagues and friends to nominate individuals that they thought had most significantly impacted the world in a positive way in recent history. This approach was similar to the method used by Matta (2004) in his research for a master's degree thesis. Names that were mentioned included former US president Bill Clinton, former US President Ronald Reagan, former British Prime Minister Tony Blair, Civil Rights leader Martin Luther King Jr., South African freedom fighter and former president Nelson Mandela, Pope John Paul II, Mother Teresa, former US Vice-President Al Gore, and Mahtma Gadhi. Based on historical name recognition, I decided to begin with Martin Luther King Jr. and Nelson Mandela. I searched a variety of databases in the University of South Dakota Lommein Health Sciences Library, looking specifically for any biographical literature that would provide information about the lives of these individuals. I also searched for speeches made by these individuals that would give me insight regarding their lived experiences.

The databases that I searched included the Proquest, EBSCO Mega FILE, Ovid, and CINAHL. Key phrases used in the search included "Biography of Martin Luther King Jr." and "Biography of Nelson Mandela" respectively. A number of articles were identified. Most of them discussed reviews of books presenting the biographies of those individuals. Some of the articles summarized the biographies.

In addition, I searched the World Wide Web using the google search engine. It quickly became apparent that one of the characteristics the two individuals shared was that they were both Nobel Peace Laureates. I therefore searched the Nobel Prize website (Nobelprize.org) for more information. The findings of the search indicated that many of the individuals who had made a significant impact in the world had received Nobel Peace Prizes. I included the award of a Nobel Peace Prize as one of my criteria for inclusion of individuals in the study. Mother Teresa of Calcutta was chosen as a participant in the study based on this criterion and the fact that she was an easily recognizable name for her work with the poor. A number of other Nobel Laureates (scientists, philanthropists, humanitarians, political leaders etc.) were identified. However, none of their names had attained the level of recognition of Nelson Mandela or Martin Luther King Jr. The biographies of many of them were summarized in articles by Abrams (1997) and Stenersen (2004). These biographies were also analyzed.

In addition to the three main Nobel Laureates (Martin Luther King Jr., Nelson Mandela, and Mother Teresa) and other Laureates included in the biographical summaries by Abrams and Stenersen, I included Pope John Paul II. The decision to include the Pope, even though he was not a Nobel Laureate was based on the fact that he was one of the most recognized individuals in the 20th century. "He was…the first non-Italian pope in 455 years…" (Cable News Network [CNN], 2000a, p. 4 of 5) and was named "Man of the year" in 1994 by Time Magazine in which it was noted that "he generates an electricity 'unmatched by anyone else'" (CNN, 2000b, p. 2 of 8). Furthermore, not only was he the most traveled pope in the history of the world, but as CNN noted, "It is doubtful there has ever been a pope who has so successfully translated his strength, determination and faith into such widespread respect and goodwill" (2000b, p. 7 of 8). He was credited, together with the former US president Ronald Reagan, with bringing down communism in the former Soviet Union and Poland (CNN, 2005a). It was clear therefore, that Pope John Paul II had been probably the most influential person in the world in the 20th century and my analysis would not be complete without him.

After this extensive search, the final sample consisted of four key individuals (Martin Luther King Jr., Nelson Mandela, Mother Teresa, and Pope John Paul II), and biographies of 13 Nobel Laureates as summarized by Abrams (1997) and Stenersen (2004), namely, Baroness Bertha von Suttner, Jane Addams, Emily Green Balch, Betty Williams, Mairead Corrigan, Alva Myrdall, Aung Sang Suu Kyi, Rigoberta Menchu Tum, Fridjof Nansen, Albert Schweitzer, Georges Pire, John Boyd Orr, and Norman Borlaug.

Procedures

The data consisting of literature identified in the search which comprised the biographical information and the key speeches of the four individuals in the sample (King, Mandela, Mother Teresa, and Pope John Paul II) were retrieved. Biographical briefs of the Nobel Laureates as summarized by Abrams (1997) and Stenersen (2004) were also printed out.

Data Analysis

The written speeches were analyzed using phenomenological methods. These included describing the phenomenon of interest in detail, reading each speech to establish a sense of what the individual was saying about the phenomenon, immersing myself in the data by reading the speeches line by line and identifying essences of the phenomenon (these included statements by individuals in their speeches indicative of their beliefs, opinions, attitudes, and values), developing thematic descriptions, and articulating a formal description of the phenomenon as perceived by the four individuals in the study (Collaizi, 1978; Giorgi, 1985; Patterson & Zderad, 1976; Streubert, 1991; van Kaam, 1959; van Manen, 1984). The phenomenon was defined as follows:

> The phenomenon of interest is the individuals' views and opinions about humanity, the world, perception of right and wrong and human responsibility in the world. Essences indicative of the phenomenon will include statements by individuals regarding what is right or wrong in their perspective, what ought or ought not to be done, what constitutes a good life, and a vision of what humanity should aspire to achieve.

The biographical information was analyzed using the heuristic method as proposed by Moustakas (1990) including initial engagement, immersion into the topic of interest, incubation, illumination, and synthesis. Similar to Matta (2004), the phenomenon of interest that I sought to understand through analysis of the biographies was the experiences of individuals included in the study that led them to adopt a particular outlook in the world. My engagement in this topic included my experiences as a poor child in Africa who at an early age developed a sharp sense of the unfairness of socio-economic institutions that kept some people poor and caused them much suffering and humiliation. That experience became the basis of my consciousness about the necessity to contribute to the best of my ability to changing circumstances in the world for the sake of fairness for all humans.

Since High School, I have immersed myself in the topic of fairness through reading and reflecting on the themes of poverty, fairness of distribution of wealth, and socio-economic institutions that contribute towards keeping some people poor and subservient to others. As such, I have had many years to think about the issues during which time there has been a period of incubation of the information related to the phenomenon. Therefore, in this analysis, my intention was to reflect on the biographies of the individuals in the study with a view to obtaining further illumination of the issue at hand and hopefully gaining insights that would lead to a synthesis of how individuals can act in their occupational pursuits to impact world issues of concern to humanity in a significant way. Therefore, based on the heuristic framework, informed by my own experiences as a person who desires to cause positive change in the world, I sought to identify experiences that energized the individuals who constituted the sample in the study and informed the visions that guided their actions that led to a positive impact in the world. I hoped that uncovering such experiences would help bring an understanding of how each person can act insightfully as a change agent through occupational performance.

Establishing Trustworthiness of the Study

Lincoln and Guba (1985) developed criteria for ensuring rigor in qualitative research. They asserted that a qualitative study should be conducted in such a way that credibility, transferability, dependability, and confirmability are assured. Credibility is attained by prolonged stay in the field, engaging participants in the research endeavor, triangulation (where data are elicited from multiple sources and examined to find out if they lead to convergence of conclusions), member checks (where research participants are contacted to confirm the researcher's interpretation of data), and peer debriefing (where a panel of researchers complete data analysis independently and debate the emerging themes). For this study, I could not engage participants since the data constituted published speeches and biographical information and I had no direct access to the research participants. For the same reason, I could not do member checks. Since I was the sole investigator, I could not do peer debriefing either. However, by analyzing both the speeches and biographical information, I was able to see convergence of findings from the two data sources (triangulation of sources) which strengthened the findings.

Transferability is achieved by thick description of the data and procedures to enable other researchers to replicate the study. This criterion was achieved in the present study. The data and procedures were described in sufficient detail so that any other investigator wishing to replicate the study could follow the outlined steps and arrive at comparable conclusions about the phenomenon of interest.

Dependability is achieved by ensuring credibility as described above. Finally, confirmability is achieved by making clear the processes that are followed to arrive at the conclusions so that another researcher can follow the logic used to arrive at similar or divergent conclusions. For this study, consistent with the process of reflexivity required in phenomenological and heuristic type research, I kept a journal where I wrote my reflections about the phenomenon as insights occurred to me as a result of immersion in the data. I also wrote memos (which provided audit trails) within the data as I interpreted them so as to keep track of the logic leading to my interpretations.

Findings

Data analysis as described above indicated that the actions of individuals included in the study were motivated by a purpose informed by a vision whose foundation was a deep love for humanity. This love made them sensitive to the suffering of human beings. This sensitivity was labeled **empathy**. Being sensitive also made them conscious of injustice and other conditions that demeaned human beings by depriving them of freedom and dignity. They made a conscious effort to enlighten themselves so that when they acted, their actions were based on a good understanding of themselves, other people, and the historical and contemporary contexts within which they were called upon to act. The above findings were expressed through four themes: **purpose-based actions, love for humanity, moral sensitivity**, and **self enlightenment**. Each of the four themes will be discussed in detail. Exemplars (direct quotes from the speeches and biographical information) will be used to ground the themes.

Purpose-Based Actions

The individuals who constituted the sample for the study lived their lives by acting in accordance with a purpose that transcended their personal self-interests. This purpose was consistent with a clear vision of a world in which humans were conceptualized to live freely, with dignity, and in harmony. For

Martin Luther King Jr., the vision was articulated in form of a dream as indicated in his famous speech: "I have a dream that one day on the red hills of Georgia, the sons of former slaves and the sons of former slave owners will be able to sit down together at the table of brotherhood" (King, 1963a, p. 3). In this statement, King's vision of a world in which human beings lived freely, with dignity, as brothers and sisters irrespective of the color of their skin was evident. Similarly, in his acceptance speech for the Noble Peace Prize, Mandela (1993, p. 2 of 4) expressed his vision of a new South Africa that was committed to:

> ...the relentless pursuit of the purposes defined in the World Declaration on the Survival, Protection and Development of Children. The reward of which we have spoken will and must also be measured by the happiness and welfare of the mothers and fathers of these children, who must walk the earth without fear of being robbed, killed for political or material profit, or spat upon because they are beggars. They too must be relieved of the heavy burden of despair which they carry in their hearts, born of hunger, homelessness and unemployment.

This vision, as is evident in the above two quotes, was grounded on a deep love for humanity. It was also based on their own backgrounds and experiences which included religious beliefs, upbringing, experience of adversity, etc. For example, Mother Teresa's vision consisted of a desire to serve the poor by making them feel wanted and loved, and reducing their sense of isolation. The basis of this vision was a "basic philosophy of life" which was "firmly rooted in her Christian faith" (Norwegian Nobel Committee, 1979, p. 1 of 2). Similarly Pope John Paul II was raised in a family that was strictly Catholic but who "did not share the anti-Semitic views of many Poles. One of Lolek's (Pope John Paul II's name before he became Pope) playmates was Jerzy Kluger, a Jew who many years later would play a key role as a go-between for John Paul II and Israeli officials when the Vatican extended long-overdue diplomatic recognition to Israel" (Christensen, 2000a, pp. 1-2 of 4). This early exposure to diversity helped explain his vision of a world of dialogue and brotherly love where all human beings lived in freedom and dignity.

These experiences made the individuals concerned develop a deep empathy for other human beings. This empathy was best demonstrated by Nelson Mandela who tried to understand the position of his oppressors while he was in prison at the notorious Roben Island:

In particular, he studied Afrikaans and learnt to understand the mind-set of the Boer (the white Dutch who were the colonizers during Apartheid South Africa) minority through discussion with warders and staff. Mandela identified with these descendants of Dutch seventeenth-century immigrants and saw that he himself, under other circumstances, could have taken views similar to theirs. And he always appreciated their fight against the English in the Boer war...

Guided by this vision, these individuals were ready to act (perform their occupations) with commitment and determination, sometimes even in the face of danger to themselves, to right the wrongs that they saw in the world. This determination was best illustrated by Pope John Paul II (2005, p. 4 of 5) in his assertion that: "Bringing about an authentic and lasting peace in this violence-filled world calls for a power of good that does not shrink before difficulties." Recognition of the need and readiness to act was also illustrated by statements such as the following: "Love begins at home, and it is not how much we do, but how much love we put in action that we do" (Mother Teresa, 1979, p. 3 of 5). In other words, Mother Teresa recognized that professing love for people without action did not amount to much.

Finally, in their readiness to act, these outstanding people recognized that global problems were a consequence of wrong human actions or lack of actions. For example, King (1963b, p. 5) categorically stated that:

We will have to repent in this generation not merely for the hateful words and actions of the bad people but for the appalling silence of the good people. Human progress never rolls in the wheels of inevitability; it comes through the tireless efforts of men willing to be co-workers with God, and without this hard work, time itself becomes an ally of the forces of social stagnation.

The above statement seems even more pertinent in the context of the argument being advanced in this book that in order to change global issues that are problematic, we have to act consciously as individuals through our daily occupations in such a way as to influence issues positively. In a similar vein, Mandela (2005, p. 2 of 3) stated that: "Like slavery and apartheid, poverty is not natural. It is manmade and it can be overcome and eradicated by the actions of human beings."

Finally, these individuals acted in the context of the occupational roles that society had assigned to them. However, informed by their visions, they went beyond the confines of their roles and addressed wider societal issues. King launched his Civil Rights work from the pulpit and used his skills as a preacher (his main productive occupation) to perform this more global role. He was not content to assume a detached posture by preaching the Bible from the safety of the Church. Similarly, Pope John Paul II acted in the context of his role as a Pope but as was observed by the CNN (2000a, p. 3 of 8): "Not content with tending merely to church affairs, John Paul has made the world his business - especially in regard to human rights." Boyd, the 1949 Nobel Peace Prize Laureate was a nutritionist who saw his role as going beyond just serving as a nutritional consultant as was expected. Instead, he went beyond those expectations and served a wider vision of providing "better nutrition in a wider context..." (Sternersen, 2004, p. 6 0f 11). Many more similar examples can be cited.

Love for Humanity

As mentioned earlier, all the individuals included in the study demonstrated a profound love for humanity. This love was based on deep empathy which made them sensitive to the suffering of other human beings. It was illustrated in the biography of Mother Teresa who:

> ...had a glimpse of the poverty and squalor of the slums, of sick people who remained untended, of lonely men and women lying down to die on the pavement, of the thousands of orphaned children wandering around with no one to care for them. (Nobel Foundation, 1979c, p. 2 of 5)

It was her empathy for these people that made her sensitive to their suffering and loneliness and became the basis of her vision. Pope John Paul II (2005, p. 1 of 5) expressed a similar capacity for empathy in his statement while addressing the diplomatic corps accredited to the Holy See. He stated:

> These sentiments of joy (referring to Christmas sentiments) are overshadowed, unfortunately, by the enormous catastrophe which on 26 December struck different countries of Southeast Asia and as far as the coasts of East Africa (he was referring to the Tsunami that originated in Indonesia and caused hundreds of thousands of deaths and devastation). It made for a painful ending of the year just past...

In this statement, one can clearly see the pope's ability to identify with the pain of those who had suffered the devastation.

The deep empathy described above emanated from these individuals' sense of connectedness with all humanity, and their recognition that inclusiveness held more hope than separation and isolation. This connectedness was expressed by King (1963a) in his assertion that any thing that affected a single member of the human species affected us all. Specifically, he stated: "Injustice anywhere is a threat to justice everywhere. We are caught in an inescapable network of mutuality, tied in a single garment of destiny. Whatever affects one directly affects all indirectly" (p. 1). This notion of interconnectedness not only between all humans but also between humans and all other living things was emphasized in occupational science by do Rozario (1997) and Wilcock (2006). It has also been propounded by various mystical philosophical systems such as Taoism, Shamanism, and Hinduism, and recently by modern scientific disciplines including physics and the life sciences (Capra, 1999, 1996).

The proposed sense of interconnectedness implies that we cannot dismiss anything that happens to fellow human beings as irrelevant. We cannot say that the misfortunes of others are "not our business". We are individually and collectively responsible for what happens in our world. Mandela (1994) articulated this very well in his statement that: "We as an individual within the species of mankind, must realize that we are our brother's keeper and have a duty to ourselves to help others grow and flourish and by doing so, we emulate the power within that truly makes us man the powerful" (p. 2 of 4). This quality is recognized by the Norwegian Nobel Committee which chooses for the Nobel Peace Prize "those who are 'champions of brotherly love' or 'self-sacrificing'" (Stenersen, 2004, p. 9 of 11).

In the theme of love for humanity, the attitude of acceptance of other people without judgment was also prevalent. King (1963a, p. 2) demonstrated this unqualified acceptance in his statement that: "The marvelous new militancy which has engulfed the Negro community must not lead us to a distrust of white people, for many of our white brothers, as evidenced by their presence here today, have come to realize that their destiny is tied up with our destiny." In other words, King did not want to judge white people merely because those who had oppressed him and those of his race were white. He was ready to accept and work with all those who shared in his vision. Mother Teresa learned this non-judgmental attitude in a special way from one of his clients, a homeless man who was dying of starvation and disease. Her contact with this man was very moving as indicated by the following statement: "And it was so wonderful to see the greatness of that man who could speak like that, who could die like that without blaming anybody, without cursing anybody, without comparing anything" (Mother Teresa, 1979, p. 3 of 5).

This non-judgmental acceptance of all people made these great people willing to forgive easily any transgressions against them for the sake of reconciliation. For example, Mandela (1994, p. 1 of 3) stated: "The time for the healing of the wounds (following the apartheid period in which black Africans were brutally oppressed, and some like Mandela himself arbitrarily imprisoned and many even killed) has come. The moment to bridge the chasms that divide us has come. We must therefore act together as a united people, for national reconciliation, for nation building." Similarly, when Pope John Paul II was shot and almost killed, "It didn't matter to the pope who was responsible, and later he visited Agca (the man who shot him) in his cell and forgave him" (CNN, 2000a, p. 4 of 8).

Aung San Suu Kyi, a human rights activist in Burma and recipient of the Nobel Peace Prize in 1991 was recognized for her "commitment and tenacity with a vision in which the end and the means form a single unit. Its most important elements are: democracy, respect for human rights, *reconciliation between groups*, non-violence, and personal and collective discipline" (Abrams, 1997, p. 9 of 15, emphasis mine). In other words, these individuals resisted the temptation to pursue vengeful justice for the wrongs committed against them. Instead, they favored forgiveness and reconciliation for the sake of human well-being.

This forgiveness and readiness to reconcile made them seek dialogue even with those who did not share their views. Pope John Paul II was particularly emphatic about the need for dialogue as a way to resolve differences for the sake of human welfare. He once stated that: "To the extent that modern world stifles dialogue among cultures, it heads towards conflicts which run the risk of being fatal for the future of human civilization" (Pope John Paul II, 1983, p. 3 of 5). In other words, he saw dialogue as essential for the sake of survival of humanity.

Because of the love for humanity, these individuals were averse to any actions that could lead to destruction of life. Martin Luther King Jr. insisted that the struggle for civil rights had to be carried out without violence, even in the face of the threat of violence from the authorities. As Burrow (2006, p. 298) stated: "To the end, he resisted incitement to violence, and tribal retreat". Similarly, Mandela was committed to non-violence in his struggle for South African Independence until there was no other alternative. This was clear in his statement from the dock in 1964 when he was on trial for his life under the accusation of sabotage against the state, a crime punishable by death.

Disputing the allegation that members of the "Umkhonto we Sizwe (Spear of the Nation)", the armed wing of his political party [the African National Congress (ANC)] were involved in violence, he explained that "...strict instructions were given to its members right from the start, that on no account were they to injure or kill people in planning or carrying out operations" (Mandela, 1964, p. 8 of 19). He continued to explain that their mission was to sabotage state infrastructure and cause economic hardships until there was enough political support by the public to force the government to change its apartheid policies. Thus: "...Umkhonto members were forbidden ever to go armed into operation" (p. 8 of 19). Pope John Paul II's views were consistent with those of King and Mandela. As Christensen (2000b, p. 5 of 9) asserted, "The pope is very negative about destroying a human life to save another human life (referring to embryonic stem cell research)". This sentiment recurred among many of the Nobel Peace Laureates whose biographies were analyzed. For example, The Nobel Committee appreciated Schweitzer's (one of the Laureates) "cultural philosophy, 'Reverence for Life'" (Stenersen, 2004, p. 4 of 11).

Because of this reverence for life, these great individuals believed that all people of good will should work together to create conditions that would make it possible for every individual on earth to have justice, live with dignity, and realize his/her full potential as a human being. This belief was best articulated by Pope John Paul II in his statement that:

> All together, we can effectively contribute to respect for life, to safeguarding the dignity of the human person and his or her inalienable rights, to social justice and to the preservation of the environment....As Christians, we cannot keep silent and we must denounce this cultural oppression which prevents people and ethnic groups from being themselves in conformity with their profound vocation. (Pope John Paul II, 2004, p. 3 of 4)

Moral Sensitivity

Because of their sense of connectedness and love for humanity, the individuals in the study had a moral conviction that was expressed as consciousness of a sense of justice, including just distribution of material goods which they saw as a necessary condition for preservation of human dignity. It was on the basis of this moral sensitivity (ability to identify morally repugnant circumstances [Sadler, 2004]) that both Martin Luther King Jr. and Nelson Mandela found it necessary to disobey some of the laws that were unjust. King (1963b, p. 3) said:

> There are two types of laws: just and unjust. I would be the first to advocate obeying just laws. One has not only a legal but a moral responsibility to obey just laws. Conversely, one has a moral responsibility to disobey unjust laws.

They saw just laws as constituting a framework within which human beings could live freely and in dignity. This meant that just laws also regulated just distribution of material goods so that everybody had the necessities to make living with dignity possible. That is why Mandela (2005, p. 2 of 3) expressed the opinion that:

> ...overcoming poverty is not a gesture of charity. It is an act of justice. It is the protection of a fundamental human right, the right to dignity and a decent life... The steps that are needed from the developed nations are clear. The first is ensuring trade justice.

Similarly, Pope John Paul II (2005) was of the opinion that: "While this principle (of universal destination of the earth's goods) cannot be used to justify collectivist forms of economic policy (he was against communism because he thought it was repugnant to human freedom and dignity), it should serve to advance a radical commitment to justice and a more attentive and determined display of solidarity" (p. 3 of 5). Out of this moral sensitivity, these individuals were concerned about global issues since justice and human dignity necessarily meant a concern about circumstances in the world that threatened those values by bringing about injustice, human suffering, and threat to freedom.

Self-Enlightenment

The final theme that emerged in the study was self-enlightenment. The individuals whose biographies and speeches were analyzed developed their visions and moral convictions that guided their actions by first consciously electing to learn about the issues of concern in the world.

In other words, they learned about the historical and contemporary context within which they were called upon to act. This consciousness about the world was best articulated by Pope John Paul II (1994) as follows:

> It would be appropriate to seek historical, cultural, social and intellectual causes (of secularization of social and private life), and at the same time to promote a respectful and open dialogue with those who do not believe in God or who profess no religion...(p. 1 of 3)

He therefore sought and encouraged his clergy to make attempts to understand the historical, social, cultural, and intellectual factors that led many people to push religion out of their lives. In addition to understanding the external circumstances that caused events in the world, they also sought to increase their self-understanding, so that when they acted, they were able to do so with authenticity and confidence. This devotion to self-knowledge was evident in the biography of Nelson Mandela which indicated "a man who over nearly 40 years in prison, thought deeply," reached "much and tried his best to understand human nature" (Mark, 2002, p. 1 of 4). Through this deep reflection, "he truly understood that the light of God's path brought forth much in the way of inner peace and sanctuary and the total power of learning to be oneself" (p. 2 of 4).

Discussion

One of the findings in this study was that the individuals whose biographies and speeches were analyzed tended to act with purpose. This purpose was based on a vision which was grounded on a deep love for humanity which emanated from empathy for the suffering of other human beings. Because they had a vision, they were able to launch their actions from the context of the occupational roles assigned to them by society, but they transcended the constraints of those roles to address the more global issues important to humankind.

The above finding implies that people who want to do something to change circumstances in the world must first reflect and articulate a vision to guide their actions. Guided by this vision, they can act authoritatively without feeling constrained by limitations imposed by the occupational roles assigned to them by society. However, the visions of the great people who were the subjects of this study were not ego driven. They did not focus on achievements for personal glorification or gain. Rather, the visions were connected to what Dyer (2004) called the "intention", or the creative energy (which he equated with God) in each one of us. Dyer postulated that while the ego part of you "believes you're separate from others, separate from what you'd like to accomplish or have, and separate from God" (p. 39), part of the intention was imagination which conceptualized everybody and everything as connected to everybody and everything else, and to God.

It is this imaginative energy that these individuals used to create a vision that guided their actions. That is why other related themes that emerged included love for humanity which was prompted by connectedness and empathy, and moral sensitivity based on consciousness about the need for justice, with the recognition that injustice that caused suffering, loss of freedom and dignity for one person constituted suffering, loss of freedom and dignity for us all. Therefore, these individuals were not ego driven but rather their visions came from "the intention" as indicated by the fact that they possessed some of its characteristics as proposed by Dyer. Some of these characteristics included: creativity (they conceptualized novel solutions to existing problems, such as non-violent resistance through civil disobedience which was the weapon of choice by Martin Luther King Jr. in his struggle); kindness; love (they all loved humanity and were averse to anything that would cause loss of life); valuing of beauty; and receptivity (they were open to understanding others, including those who had different views and lifestyles, and they were willing to have a dialogue with them).

Therefore, these individuals distinguished themselves by committing to a process of searching for enlightenment that would lead to becoming what was referred to in positive psychology as "an optimal functioning person, to living 'the good life'" (Matta, 2004, p. 5). They were able to make this journey towards optimal functioning by being in touch with their inner intention as defined by Dyer (2004) and from this depth, to create "a visionary goal in mind - aspiring towards greatness" (Matta, p. 5). In essence, the findings of this study indicated that these individuals achieved what Maslow (1970) referred to as self-actualization characterized by a sense of novelty; confidence (they were comfortable with whom they were and did not rely too much on approval by others); non-prejudicial attitude (as indicated by the finding that the participants in the study were non-judgmental and accepting of others); and empathy (they were sensitive to the suffering of humanity).

Matta (2004) proposed a hypothesis that the traits of great individuals as found in the present study and discussed above were not random chance occurrences. Rather, they were "at least partially a result of who they are…" (p. 4). This could be mistakenly taken to imply that such traits could not be developed. Some people had them but the majority of us didn't. Therefore, some people could impact the world significantly because they were destined to be great while others would not have much effect. The contention in this book is that the above possible implication of Matta's hypothesis is not correct. Anyone can consciously develop the traits demonstrated by individuals who were the subjects of this study and therefore can contribute towards changing issues of importance in the world and therefore become great. That is why the study was entitled "Ordinary people who do great things" to emphasize the fact that anyone can achieve greatness if he/she chooses to do so. In the rest of this chapter, ways in which individuals can develop such traits will be suggested.

Implication of the Findings for Occupation-Based Approach to Resolving Global Issues

The purpose of this chapter was to present a framework for occupation-based approach to resolution of global issues by reflective individual choices and participation in occupations in such a way that global problems can be influenced positively. In other words, the hope is that by following the guidelines provided in this chapter, individuals can be proactive in addressing global issues of concern for the benefit of all humanity and other living things on the planet. The above described study was completed so as to identify some characteristics of individuals who have influenced events in the world significantly. It was hoped that by identifying such characteristics, individuals could learn to cultivate them in themselves so that they can be equally effective.

The first characteristic identified in the study was *purpose-oriented* actions. The purpose on which actions were based was grounded on a clearly defined vision which was informed by a love for humanity and empathy for the suffering of other human beings. The individuals drew from their own backgrounds a foundation for the development of that empathy (sensitivity to the suffering of others) and therefore formulated a vision of a world in which humans lived freely, with dignity, and in harmony. This implied that anyone who wanted to impact events in the world positively needed to formulate a clear vision to guide actions that would lead to a positive effect in the world. However, the vision needed not to be ego-driven. Rather, it had to be based on identification with humanity and other forms of life, an understanding of the suffering of those life-forms, and a desire to improve their lot out of love instead of a desire to gain personally.

To formulate the vision described above, the individual may want to reflect on his or her life and identify experiences that would enable him/her to develop the empathy necessary to understand the condition of those who are in unfortunate circumstances. For example, if you have never lived in abject poverty, you may not understand how it is to live under those conditions. However, you may understand that poverty leads to feelings of humiliation and a sense of isolation (due to a feeling that no one cares). Even though you may never have lived in abject poverty, you probably have at one time experienced the feelings of dejection, isolation, or humiliation. It may have been a romantic partner who rejected you. It might have been a group of peers at school or work who did not want to associate with you. These events made you feel dejected, isolated, and probably humiliated. By drawing upon these experiences, you can begin to understand, in a small way, the experiences of those who live in hopeless conditions due to poverty. This could be the basis for developing feelings of connectedness and empathy, and a love for humanity that could inform your vision which would in turn guide your subsequent life choices.

The second characteristic that was found to be prevalent among individuals in my study was a *love for humanity*. As mentioned above, this love was based on the ability to feel connected and empathic towards other human beings. It has already been discussed above how one could make efforts to develop the sense of connectedness and empathy by drawing from personal experiences.

The third characteristic that was common to all the individuals included in the study was *moral sensitivity*. By moral sensitivity was not meant adherence to morality in the abstract and narrow sense as some people seemed to conceive religious morality. Rather, the moral consciousness being referred to here was development of a deep sense of justice grounded on the love for humanity mentioned earlier. It was sensitivity to any circumstances that caused suffering, loss of freedom, humiliation, and loss of dignity for a fellow human being. This necessarily included a concern for equitable distribution of the world's goods to enable all human beings to live freely and in a dignified manner.

For example, one cannot have moral sensitivity and yet be unconcerned about the fact that some people live on the streets, eating out of garbage cans and begging for cramps of food just to stay alive. Such a life is not dignified and therefore it is not morally right that some people are forced to live under those conditions. The above finding implies that if one wants to make choices so as to participate in occupations in such a way as to influence events in the world positively, he/she may want to develop moral sensitivity necessary to distinguish between wrong and right actions from the point of view of how those actions influence issues that affect the quality of life of humans and other living things, rather than from an abstract, narrow interpretation of the Bible or some other religious book.

The other finding in my study was that individuals who constituted the sample made conscious *efforts to enlighten themselves*. Through reflection, they sought to understand themselves so that they could act authentically and with confidence. They also sought to understand the world so that they were clear what it was that they hoped to change by their actions. This finding implied that in order to contribute effectively to resolution of global issues of concern, one needed to make a conscious effort to become informed about those issues and the historical and contemporary contexts that made their resolution challenging. This conclusion is consistent with the finding in an earlier study reported in Chapter five that education significantly influenced readiness to change occupational performance patterns for the benefit of the earth's ecology (Ikiugu et al., 2007).

This book was an attempt, in a small way, to help you acquire the information that you need in order to be a proactive change agent for the sake of Gaia. By reading chapters one through six of the book, you are now aware of some of the issues of concern. The information presented in those chapters can be a point from which you can launch your intellectual curiosity in order to learn more about the world, issues that threaten humanity and other life forms, and methods that have been proposed to resolve them. This would help you become *virtuous through enlightenment* as proposed by Socrates (see discussion in chapter 6).

Finally, each of the individuals who were the subjects of the study recognized the need to act in order to change circumstances in the world. A vision not supported by action could be seen as mere fantasy. Therefore, if you want to change something in the world, you have to make a decision to act. However, such action needs to be informed by the ideals of the vision that you articulate to guide your life as described above. Furthermore, you do not have to feel that your ability to act is limited by the social position that you hold based on your occupational role. If you have a vision, you can find a way to act creatively while performing your role, whether that role is a parent, teacher, preacher, cleaner, etc. It is suggested that one way of changing events in the world is by paying attention to what choices you make and how you perform occupations in the process of fulfilling your roles as a worker, family person, friend, lover, social participant, political participant, as well as in your leisure pursuits and self care.

While emphasizing individual action through occupational performance as a way of resolving global issues, it is recognized, as mentioned earlier, that not all problems can be addressed through individual action. Systemic approaches are very important. For example, political solutions are necessary in order to solve the problem of environmental pollution through legislation. Similarly, legislation and establishment of social policy by governments are the two variables that may be the most effective in regulating distribution of material goods in societies. That is why governments have to commit to confronting some of the major global issues such as environmental destruction and global warming/climate change, material inequalities, corruption, etc. Even education of individuals so that they are able to think critically in order to make daily decisions to act as proposed in this model requires systemic regulation of school curricula (e.g. by introducing ethics as a subject of study early in school curricula) as mentioned earlier in chapter 6 [see also the study findings by Ikiugu et al. (2007) reported in chapter 5].

However, even in those cases, individuals have a large role to play. It is individuals who elect politicians into office. Therefore, by paying attention to the kind of people you choose to represent you in government, you determine the policies that the government puts in place and how those policies affect the global issues in question. Through the ballot, you have the power to bring pressure to bear on your government to establish appropriate policies as urged by Gore and colleagues in the pledge for the Live Earth

(2007) initiative. In other words, individuals cannot be excused when issues of concern are not resolved. As Eakman (2007) asserts, the way you interact as an individual with the system in which you live determines to a large extent the kind of culture that emerges and whether such culture affects the issues in question positively or negatively.

In addition, other activities of individuals directly affect the global issues in question. As an illustration, the means of transportation that you choose to use directly affects the issue of environmental pollution. As illustrated in chapter 5, your personal attitude towards material wealth affects how you choose to acquire and accumulate things and whether in the process of seeking wealth for yourself, you deny other people the opportunity to acquire a means of their own sustenance. Fortunately, as Ikiugu et al. (2007) found in their study, people are becoming increasingly aware of their responsibility in influencing global issues positively or negatively through their occupational choices and performance patterns. We may therefore be on the verge of a cultural paradigm shift for the benefit of all life on earth.

The cultural paradigm shift postulated above is an essential necessity. Through the discussion in previous chapters, I hope that I have conveyed the urgency with which we, as a human race, need to confront global issues that face us today. We have to make a decision whether we want to continue living as a species or whether we are ready to go extinct. If we choose to live, then we have to do the following immediately: 1) reduce emission of greenhouse gasses to a minimum (stopping them altogether might actually be the only viable option according to Lovelock); 2) stop polluting the environment; 3) stop cultivating any more land for food production so that the earth has a chance to heal itself from the wounds that we have inflicted on her; and 4) curb population growth (we actually need to reduce World population because right now, the earth's resources are stretched to the point of being overwhelmed). For the above four objectives to be realized, it is necessary to eliminate corruption and other functional maladies afflicting governments and other social institutions, eliminate abject poverty, and educate all earth's citizens (educated individuals who are not poor are more empowered to protect the environment). Failure to do the above is a vote for elimination of not only humanity but all other living things on the planet earth.

Even though governing institutions are largely responsible for achieving the above enumerated objectives, individuals can do their part. The framework being presented here is meant to capitalize on the increasing awareness of individuals of their responsibilities to the earth and all life on it by guiding them to make choices such that their actions contribute significantly to resolution of the global issues of concern. The framework will be informed by the findings of the study presented earlier in this chapter regarding the characteristics of individuals who have significantly impacted the world positively.

Framework for Individual, Occupation-Based Approach to Resolving Global Issues

It is proposed that a conceptual model of practice based on pragmatic principles which was developed for use in occupational therapy (Ikiugu, 2004a, 2004b, 2004c, 2007) be used to guide individuals in learning how to consciously make choices and participate in occupations in such a way as to influence global issues of concern positively. The model consists of three phases: Phase I - belief establishment; the individual establishes a mission statement defining the legacy that he/she would like to leave in the world, identifies occupations whose regular performance would lead to achievement of that mission, and rates himself/herself on four scales (frequency, adequacy, satisfaction with performance of each of the occupations, and belief about ability to perform the occupation with desired frequency and adequacy), Phase II - action; the individual establishes goals to facilitate performance of each of the identified occupations to his/her satisfaction, and Phase III - consequence appraisal; progress towards the established goals is evaluated on a regular basis. Based on the findings of this regular re-assessment, decisions are made about whether to adjust actions, beliefs, or both actions and beliefs.

The above described conceptual model will be modified for application to assist individuals evaluate their occupational performance patterns and change their lifestyles as necessary to increase chances of impacting global issues of concern positively. In this section, steps of the suggested framework for assisting individuals change their occupational performance will be presented as follows: individual education as a foundation for belief change; personal life mission establishment and self-assessment; and action.

Individual Education as a Foundation for Belief Change

The framework presented in this chapter is meant to empower you. The premise is that although you may think that you are only a single individual, who may not have much influence in the world, you are still very powerful because what you do has a disproportionately large impact in the world, either for better or for worse. You therefore need to think about what you do every moment and the effect of these activities not only in your life but in the lives of other human beings, your community, environment, the lives of other living things, and the earth's ecosystem at large.

In order to act with this sense of responsibility, it is essential that you understand what other people and other living things may be experiencing in this world. Like the great individuals discussed in the study presented earlier, you need to feel a sense of connectedness and empathy for other human beings and life-forms. Around the mid-20th century, Carl Rogers, the founder of professional counseling and one of the greatest psychotherapists in recent history, introduced the concept of empathy (Rogers, 1950, 1961). This referred to the ability to perceive reality through another person's frame of reference, as it was shaped by his/her experiences. This capacity is what makes us human. Using our own experiences as a frame of reference (see the discussion section in the study presented earlier), we know how it feels to be in pain, humiliated, joyous, in love, etc. We can have a certain understanding of other peoples' perspectives when they are in similar circumstances.

It is proposed that in order for you as an individual to be able to adjust your occupational choices and actions in such a way as to positively impact global issues such as poverty, material inequalities, injustices resulting from corruption and failure of governments/social institutions, etc., it is necessary to develop empathy for individuals affected by these issues. One way of developing such empathy is to expose yourself to circumstances through which you can share the experiences of people affected by those issues. I will illustrate this with a personal example. My background, as mentioned earlier, consists of growing up in a poor family where the prospects for any kind of success in life were almost negligible. One of the teachers in primary school once suggested to me that based on my family's economic status and my behavior in school (evidently I was not very compliant with institutional rules), I was not going to amount to anything more than a chicken thief (apparently not even a very respectable thief). Because of those experiences,

I can empathize with people who live in abject poverty. I can understand such individuals' sense of humiliation, and how they may give up trying to improve their lives because it seems as if there is nothing they can possibly do to make a difference.

Similarly, when I was a new occupational therapist in the United States, I remember often going to clients' rooms to treat them and they would mistake me for someone looking for a job as a cotton picker (these were older clients in nursing homes in the South). Sometimes, a client would order me out of the room and threaten to call the police if I did not leave immediately. Based on those experiences, I can empathize with minorities in certain parts of USA who assert that racism continues to exist, and limits their ability to be successful as individuals. In other words, I can understand when minorities argue that the system is unfair and odds are stacked up against them. If you face challenges such as what I endured as an occupational therapist in the South, where some individuals would not even accept my services for no other reason than my African accent and the fact that I was black, how can you compete effectively in the same career path with those who do not have to face such barriers (in my case, white therapists obviously had an advantage and were more likely than me to succeed in the occupational therapy career in the South for no other reason than the fact that they were American-born whites)? I can therefore understand why minorities require policies such as the affirmative action in order to give them at least a fighting chance in a society where many odds are still against them.

At the same time, I have been able to move more or less into the middle class and I know how frustrating it is when I work hard and a big chunk of my check goes into taxes. I can understand those in the middle and upper socio-economic classes who argue that they are being taxed too much. In my case, because of my empathy for the poor, I do not mind being taxed so that those who are in disadvantaged situations can be extended some kind of financial assistance to enable them to keep a roof over their children's heads, feed those children, clothe them, and pay for their education. I would have loved for someone to provide my family with such a break when I was a child. Nevertheless, I can understand how someone who has never been poor may find it difficult to empathize with such disadvantaged individuals.

I hope that the above discussion, taken together with the findings in our study discussed earlier, illustrates that the best way to develop the capacity to empathize with those affected by global issues of our time may be through exposure to their lifestyles. Of course, such exposure is sometimes not possible. For example, it may not be possible to live the life of a poor person for a length of time necessary to appreciate the difficulties faced by individuals who live in abject poverty. However, listening to their stories might give you significant insights regarding the realities of the world in which they live.

Similarly, no one may really know what it feels like to be a polar bear living in the Arctic Circle. There is no way to even interview a polar bear to find out what it means to the species for the polar ice to be melting, destroying their habitat. Therefore, it might be difficulty to empathize with such creatures, or with many other animal species that continue to be threatened by human destruction of forests, pollution of the environment resulting in global warming, etc. In those circumstances, the best way to develop sensitivity is to gather information about what is happening. I did not know the real impact of global warming on our ecosystem until I started reading about the phenomenon and looked at the documented statistics and stories describing the extent to which climate change is causing havoc in the lives of all living things on our planet.

In addition, we need to relate this information to our own lives. We know for instance that birds are changing their migratory patterns and many of the species are beginning to become extinct (see detailed discussion in chapters 3 and 5). This should be alarming to us because it could be a signal of the impending threat of climate change to human beings. Birds are used to test whether the constitution of gasses in mines is acceptable for human beings because they are sensitive to environmental changes (the famous canary in the coal mine). Therefore, self-interest should dictate that when other animal species start facing difficulties and their survival being threatened due to climate change, human beings should feel threatened as a species and seek to do something about it.

Therefore, the first step in attempting to do your part in meliorating global issues of concern as an individual is to educate yourself (see the theme of self-enlightenment identified in the two studies discussed earlier) and others as much as you can, about these issues. Where possible, seek exposure to experiences of those who are affected by these concerns. If that is not possible, read about the issues, attend conferences, seminars, workshops, etc. where you can obtain information about them.

Personal Mission Establishment and Self-Assessment

After educating yourself about pertinent issues facing our world, you are now in a position to make a decision regarding what you are willing to do to help make the world a better place for both humans and other living things. As a rancher/author Dan O'Brien once stated, resistance to negative influences in one's existence can only be achieved by maintaining a positive attitude (O'Brien, 2007). Furthermore, this positive attitude originates from a clear definition of one's identity (labeled self-understanding in the study reported earlier in this chapter) which provides a reason for doing what he/she does in this life. In the end, life is not about external material things, but about self-definition. In this book, it is proposed that the most effective way to create such self-definition might be by establishing a personal mission statement (vision according to the individuals whose biographies and speeches were analyzed in our study) indicating the kind of legacy that you would like to leave in the world. Such a statement is a useful tool to help you make appropriate choices and act accordingly in pursuit of daily occupations in self-maintenance, productivity, and leisure.

However, before making a personal mission statement, it may be useful to gauge your current performance in order to establish a baseline to help you assess the effectiveness of changes that you subsequently make. This is helpful because it will provide you, in a tangible way, with a means of judging how much progress you are making towards your goals, thus providing you with a sense of competence, which in itself is empowering. To help you assess your performance baseline, it is suggested that you begin by documenting in a Daily Occupational Inventory [DOI] (Ikiugu, 2007) the occupations in which you are engaged every day from 6:00 AM through midnight for 4 to 7 days (see figure 7-1 on the next page for an example of a one-day inventory).

Daily Occupational Inventory (DOI)

Please enter each occupation in which you participate next to the indicated time. For example:

6:00 am – Woke up, got out of bed, exercised, meditated, had breakfast.

7:00 am – Drove to work, listened to the news over the radio.

8:00 am – Got into the office, checked voicemail, checked email messages, answered emails, made a list of activities for the day…

Make sure to indicate the date (MM/DD/YY) for which you are logging occupations.

Name _Gerald   **Day # 1**   **Date _3/6/2006_____**

6:00 am – Combed hair, washed dishes, made coffee, ironed clothes, got dressed, made toast, got dressed, ate breakfast, watched TV, worked on computer

7:00 am – Worked on the computer, watched TV

8:00 am – Checked emails, drove to work (attended a seminar 70 miles away)

9:00 am – Driving on my way to the seminar

10:00 am – Arrived at the seminar, attended morning sessions

11:00 am – At the seminar

12:00 am – Lunch break (ate lunch)

1:00 pm – At the seminar

2:00 pm – At the seminar

3:00 pm – At the seminar

4:00 pm – At the seminar

5:00 pm – Seminar over, drove back home

6:00 pm – Driving (on my way home)

7:00 pm – Got back home, ate dinner (takeout), watched TV, worked on the computer

8:00 pm – Checked emails, surfed the Internet, worked on the computer, watched TV

Figure 7-1

Generate a list of occupations performed in the chosen time period and the frequency of performance of each occupation based on the number of times it was entered in the DOIs (see example in Figure 7-2).

List of Occupations Performed by Gerald over the 4 Days

Activity		Frequency
Activities of Daily Living		
Took a shower		4
Combed hair		4
Got dressed		4
Eating (Nutrition)		12
Went to bed		4
Instrumental Activities of Daily Living		
Mobility:	Drove to work related events - 70 miles away	4
	Drove to work related events – 20 miles away	2
	Drove to work – 5 miles away	5
	Drove to class – 4 miles away	2
Health Management: Exercised		4
Home Management: Washed dishes		4
•	Ironed clothes	4
Meal Preparation: Made toast		4
	Made coffee	4
	Made dinner	2
Communication: Checked emails		8
Education		
Attended class		4
Worked on school related tasks on the computer		19
Work		
Worked		26
Play		
Leisure		
Watched TV (for entertainment)		25
Surfed the Internet (for leisure)		5

Figure 7-2. Example of a list of occupations developed from DOIs. Instrument adapted from "Appendix B", by M. N. Ikiugu, 2007, Measuring occupational performance: A pragmatic and dynamical systems perspective. *Journal of Occupational Science,* Copyright, 2007, by the Journal of Occupational Science.

Next, follow the instructions in section I of the Modified Assessment and Intervention Instrument for Instrumentalism in Occupational Therapy [MAIIIOT] in Figure 7-3 (see original AIIIOT in Ikiugu, 2004c, 2007) to establish a personal mission statement.

Modified Assessment and Intervention Instrument for Instrumentalism in Occupational Therapy (MAIIIOT)

When using this instrument, find a quiet place where you can concentrate and respond in detail to all items without interruption. Take as much time as necessary to respond to all items in this instrument exhaustively. The instrument consists of four sections. In section I, you will create a personal mission statement. This provides a purpose towards which you will strive, including the relationship between your life and global issues of concern to humanity. In section II, you will identify occupations in whose regular engagement would lead to achievement of the stated mission in life and to meliorating global issues. In section III, rate your self-perception of engagement in identified occupations on four scales: frequency, adequacy, satisfaction, and belief in ability to engage in the occupations. In section IV, sum the rating scores to give you engagement indexes on the four scales.

I.   Personal Mission Statement

Read the following directions carefully before completing this exercise (see Covey, 1990).

Imagine that one day, you come home unexpectedly. You find a large group of people meeting in the house. As you come in, you find that for some reason, they are all talking about you. You decide to listen to what they are saying. No one knows you are there. Write down in detail what you would like to hear each of the following say about you and your role and impact on global issues such as poverty, material inequalities, environmental destruction and global warming, diseases, institutional dysfunctions, and overpopulation: (a) family member (father, mother, spouse, son/daughter, sister/brother, cousin, any other family member that you feel close to). (b) Friends (one or two close friends). (c) work/professional colleague/ associate. (d) a member of the church or some other community organization to which you are affiliated. Now, go over what you have written and take a few moments to think about what you imagine each of those people saying about you. These statements represent the kind of person you would like to be and that you can be, and your perceived destiny in regard to acting to meliorate global issues of concern. Summarize the statements in a few sentences, stating what you consider to be your personal mission statement. This mission statement will provide the direction towards which you will strive from now on. The statement should consist of four components corresponding to the four areas of the overheard conversations, and what the individuals you overheard perceived to be your responsibility to global issues of concern: family, friends, work/professional life, and engagement in Church/community organization(s).

II.   Identification of Occupations

For each of the four areas, identify two occupations in whose regular engagement will lead to achievement of the stated mission in life.

Family

1. Organizing family activities addressing broad social/global issues

_____

_____

Social Life (Friendship)

1. Discussing global issues of concern with friends

_____

2. Inviting friends to participate in programs addressing social/global issues

_____

Work/Professional Life

1. Eliminating waste at the office

---

2. Walking/riding a bike to work instead of driving

Affiliation to Church/Community Organization(s)

1. Volunteering services in programs designed to help the poor

---

2. Political participation

III. Evaluation

For each of the identified occupations, rate yourself on a scale from one (1) to four (4) regarding: (a) frequency, (b) adequacy, (c) satisfaction, and (d) belief about your ability to engage in the occupation.

Descriptors

Frequency

1 = does not engage in the occupation; 2 = rarely engages in the occupation; 3 = regularly engages in the occupation; 4 = frequently engages in the occupation as necessary.

Adequacy

1 = I am not able to engage in the occupation; 2 = engages in the occupation with difficulty and the outcome is inadequate; 3 = engages in the occupation with difficulty but the outcome is good when able to complete it; 4 = engages in the occupation, is able to complete it, and the outcome is always adequate.

Satisfaction

1= I am disappointed with my engagement in the occupation; 2 = I am somewhat satisfied with my engagement in the occupation; 3 = I am satisfied with my engagement in the occupation but would like to improve; 4 = I am happy with my engagement in the occupation as it is.

Belief

1 = I do not believe that I am capable of engaging in this occupation; 2 = I believe that I can engage in the occupation with much help; 3 = I believe I can engage in the occupation with a little help; 4 = I believe I can engage in the occupation independently with desired frequency and adequacy.

	Frequency 1 2 3 4	Adequacy 1 2 3 4	Satisfaction 1 2 3 4	Belief 1 2 3 4
Family				
1. Family activities	_X_ _ _	_X_ _ _	_X_ _ _	_ _ _ _X
2.	_ _ _ _	_ _ _ _	_ _ _ _	_ _ _ _
Social Life (Friendship)				
1. Educating friends	_X_ _ _	_X_ _ _	_X_ _ _	_ _ _ _X
2. Inviting friends to participate in activities	_X_ _ _	_X_ _ _	_X_ _ _	_ _ _ _X

	Frequency 1 2 3 4	Adequacy 1 2 3 4	Satisfaction 1 2 3 4	Belief 1 2 3 4
Work/Professional Life				
1. Eliminating wastefulness	_X _ _	_ _ _X _	_X _ _	_ _ _ _X
2. Walking/riding a bike to work	_X _ _	_X _ _	_X _ _	_ _ _ _X
Affiliation to Church/Community Organization(s)				
1. Volunteering services to the poor	_X _ _	_X _ _	_X _ _	_ _ _ _X
2. Political participation	_X _ _	_X _ _	_X _ _	_ _ _ _X
Scores (X_{11}, X_{12}, X_{13}, X_{14})	8 Frequency	9 Adequacy	7 Satisfaction	28 Belief
Scores (X_{21}, X_{22}, X_{23}, X_{24})	___	___	___	___

To obtain aggregate scores for each of the four scales, add together the ratings for each column and place the total at the bottom of the column. These scores are denoted X_{11}, X_{12}, X_{13}, and X_{14} for frequency, adequacy, satisfaction, and belief respectively.

To obtain scores during re-assessment, determine whether each of the occupations listed continue to be relevant to your mission statement. Substitute those that are no longer relevant with others that address your mission better. Rate yourself on the four scales again and aggregate the scores as you did before. Denote the re-assessment scores X_{21}, X_{22}, X_{23}, and X_{24} for the four scales respectively. Your progress in performance is indicated by; X_{21}-X_{11}, X_{22}-X_{12}, X_{23}-X_{13}, and X_{24}-X_{14} respectively.

Comments:

Figure 7-3. Modified Assessment and Intervention Instrument for Instrumentalism in Occupational Therapy (MAIIIOT). Adapted from, "Assessment and Intervention Instrument for Instrumentalism in Occupational Therapy (AIIIOT)", by M.N. Ikiugu, Instrumentalism in occupational therapy, *International Journal of Psychosocial Rehabilitation, 8*, 176-177.

As is apparent in section I of the MAIIIOT, the mission statement should encompass all aspects of your life, primarily including your family, work, socialization, and community participation. Such a statement is not only your self-definition in the sense that it provides a clear articulation of what you perceive your life to be all about, but it is also a powerful vision guiding your actions and providing a compass to give your life direction. The statement provides an indication of your perceived sense of responsibility towards pertinent global issues.

The next step is to assess your performance in relationship to your personal mission statement. This process of self-assessment is in two parts: Firstly, rank the activities in the list created from the DOIs (Figure 7-2) in order of their importance in helping you achieve your articulated mission in life, particularly with regard to helping you meet your perceived obligation to resolution of pertinent global issues. Use the top 5 listed occupations to calculate a performance score using the algorithm P_t=(SUM)P_i=SUM(PI x F), where P_t is the total performance score, P_i is the performance score for each occupation in the list, P is performance, and I is the index assigned to each occupation according to rank in the list [I=5 for activity number 1, 4 for number 2, 3 for number 3, 2 for number 4, and 1 for number 5 respectively] (Ikiugu & Rosso, 2005).

Secondly, identify occupations in the four areas of your mission statement whose regular and adequate performance would lead to achievement of the mission and a positive impact on global issues of concern (see section II of the MAIIIOT, Figure 7-3). Now, rate yourself on each of the 8 identified occupations on the four scales [Frequency, Adequacy, Satisfaction with performance, and Belief about the ability to perform the occupation with desired frequency and adequacy] (see section III of the MAIIIOT). Aggregate the scores on each scale to obtain the index for that scale. The resulting 4 indexes are denoted X_{11}, X_{12}, X_{13} and X_{14} for frequency, adequacy, satisfaction, and belief respectively.

Action

For each occupation identified in section II of the MAIIIOT whose self rating of satisfaction with performance is less than 4, establish a short-term and long-term goal to facilitate enhanced performance of the activity with desired frequency and adequacy. The long term goal is broad (e.g., by the end of one year, I will have informed at least 100 individuals either individually or in groups about the need to alter their lifestyles to reduce emission of greenhouse gasses into the atmosphere). A short-term goal should be more focused, relevant, understandable, measurable, behavioral, and attainable within a reasonable period of time (Sames, 2005).

An example of such a goal might be: Within 3 weeks, I will have changed my lifestyle significantly to reduce my contribution to accumulation of greenhouse gasses into the atmosphere as indicated by walking or riding a bicycle in every 5 out of 7 instances when I travel distances of 5 miles or less in pursuit of daily occupations. Such a goal is relevant because in pursuit of daily occupations in the areas of self-maintenance, work, and leisure, you have to be mobile, and some of the distances you travel are probably 5 miles or less; it is understandable because it is written in clear, straightforward language; it is measurable because you can count the number of times you drive distances of 5 miles or less instead of walking or riding a bicycle; it is behavioral because mobility is an observable behavior; and finally, it is attainable because if you put your mind to it, you can easily change your lifestyle to that extent in three weeks.

Note that consistent with the model illustrated in figure 5-1 in chapter 5, as an agent, your performance is largely related to the attitudes, values, opinions, and beliefs that you hold towards yourself, other people, the environment, and certain occupations. Therefore, if you find that there are occupations that you need to change in order to achieve your life mission but you are reluctant to change them, ask yourself honestly what benefits you derive from such occupations (e.g. the prestige associated with driving an

SUV). Then find other ways of attaining similar benefits without engaging in the occupation in question (e.g. maintaining your social prestige without having to drive an SUV). That is one way you can successfully make substantial changes in your occupational performance patterns.

After establishing short-term and long-term goals, the next step is committing to change your lifestyle by performing the identified occupations with required frequency. Remember that the great individuals whose lives were analyzed in our study indicated that a vision without action is sterile. Finally, set a time when you will re-assess yourself to determine whether you are making progress towards your goals. You can decide that you will re-assess yourself every 2, 3, 4 weeks, etc. During re-assessment, complete the DOIs again for 4 to 7 days, generate a list of occupations in which you engaged in that time period, rank the occupations in order of importance in helping you achieve your mission in life, and calculate your new performance score (P_{t2}).

Next, examine the 8 occupations identified in section II of the MAIIIOT and decide whether they are still essential to achievement of your life mission. If you find that some of them no longer seem to address your mission, you can substitute them with others that are more appropriate. Again rate yourself on each of the 8 occupations regarding the frequency, adequacy, satisfaction, and belief in ability to engage in the occupation with desired frequency and adequacy. You will again obtain 4 indexes representing your self-rating on the 4 scales, denoted X_{21}, X_{22}, X_{23}, and X_{24} respectively.

Your progress in performance is indicated by subtracting original scores from the current scores: $P_{t2}-P_{t1}$, $X_{21}-X_{11}$, $X_{22}-X_{12}$, $X_{23}-X_{13}$, and $X_{24}-X_{14}$ respectively. The greater the positive change in your Pt score, the more you are engaging in occupations that are essential to achievement of your life mission and therefore that are likely to have a positive impact on global issues of concern. Similarly, the greater the positive change in X scores, the more frequently, adequately, and satisfactorily you are performing occupations essential to your life mission, and the more confidence you have in your ability to perform the occupations with desired frequency and adequacy. The scores are therefore an indication of the extent to which you are progressing towards your goals and subsequently towards achieving your mission in life. Therefore, they are a source of self-empowerment. Furthermore, because of the high level of connectedness between agents in our ecological system, which is conceptualized to be a complex, dynamical, adaptive system, the effect of changing your performance is not equal to the change realized in the world. Rather, because of the recursive effect due to the feedback mechanism within the system, the change is amplified such that the outcome may be far-reaching than you may think. This will be illustrated with a case example.

Illustration: The Case of Gerald

Gerald is a cashier at a local Bank and he is attending classes in pursuit of a graduate degree at a local University. After educating himself about the issues confronting humanity, he decided that he needed to do something to be part of the solution rather than contributing to the problems. He started by documenting occupations in which he engaged for a period of 4 days (Monday through Thursday). A one day inventory of occupations in which he participated is shown in Figure 7-1. He developed a list of all the occupations in which he engaged during the four days (see Figure 7-2). Based on the MAIIIOT, he developed a personal mission statement which he articulated as follows:

I am committed to living my life in such a way that I leave this world a better place for all humanity and other living creatures. I want to live my life in such a way that when I pass on, my family, friends, colleagues, and all those who know me will say that I used all my talents, skills, and energy to: fight for fairness for all people in this world; contribute to elimination of poverty; and contribute towards saving our planet's integrity for posterity.

Gerald then ranked the occupations in Figure 7-2 according to their importance in helping him achieve the above stated mission. The top 5 ranked occupations in the list were: driving (distances of 5 miles or less), watching TV for entertainment, surfing the Internet for entertainment, exercising, and making dinner. His rationale for ranking the above occupations was: by driving distances of 5 miles or less rather than walking or riding a bike, he was unnecessarily contributing to environmental pollution and global warming; by watching entertainment TV exclusively, he was abdicating his responsibility to watch more informative programs such as congress debates or news through which he could educate himself about pertinent global issues; similarly, by surfing the Internet exclusively for entertainment, he was not optimizing use of a resource that he could employ to educate himself;

by exercising, he was keeping himself healthy (which reduced overall likelihood of incidences of diseases due to faulty lifestyle); finally, by cooking instead of eating junk food, he was contributing towards better health and less incidence of diseases.

Based on the above ranking, he calculated his P_t score as follows:

Occupation	Performance Index (PI)	Frequency (F)	P_i
1. Walking/riding a bike for distances of 5 miles or less	5	0	0
2. Watching TV for self-education	4	0	0
3. Surfing the Internet for Self-Education	3	0	0
4. Exercising	2	4	8
5. Making meals	1	10	10

$$P_t=SUM(P_i)=SUM(PI \times F)=18$$

He then identified seven occupations whose regular performance would lead to achievement of his stated mission and rated the frequency, adequacy, and satisfaction with performance for each of the occupations, and belief about his ability to perform each of them with desired frequency and adequacy. The aggregated rating scores on the 4 scales were: $X_{11} = 8$, $X_{12} = 9$, $X_{13} = 7$, and $X_{14} = 28$

Based on the above self rating, he established one long-term goal and 5 short-term goals as follows:

Long-Term Goal

Within one year, I will be regularly involved in a variety of activities alone or with family members, friends, and colleagues in order to contribute towards addressing issues of poverty; material inequalities; incidence, prevalence, and effect of diseases; corruption and poor governance; and environmental protection.

Short-Term Goals

1. Within 3 months, at least 2 members of my family and I will participate in one environmental protection activity at least once a month.

2. Within 3 weeks, I will discuss a global issue such as poverty, environmental destruction, or corruption in government with at least one friend, once a week.

3. Within 3 weeks, I will have invited at least one friend to join me in writing and sending a letter to our legislator(s) inquiring about how he/she intends to address the issue of assisting poor people to pay for their children's college education.

4. By the end of 3 weeks, I will have reduced wasteful use of paper both in the office and at home by at least 50%.

5. Within 3 weeks, I will have changed my lifestyle significantly to contribute to environmental protection as indicated by walking or riding a bike to work rather than driving 4 out of 5 times a week.

After 3 weeks, Gerald re-assessed himself to see how much progress he had made. The five top ranked occupations in his list during re-assessment were: Walking/riding a bike to work rather than driving, political participation, reducing wastefulness at the office, educating friends about global issues, and volunteering services to programs designed to help the poor. He calculated his P_{t2} score as follows:

Occupation	Performance Index (PI)	Frequency (F)	P_i
Walking/riding a bike to work	5	3	15
Political participation	4	0	0
Reducing wastefulness in the office	3	2	6
Educating friends about global issues	2	1	2
Volunteering services	1	0	0

$$P_{t2}=SUM(P_{i2})=SUM(PI_2 \times F_2)=23$$

His aggregated self-rating scores on the MAIIIOT at re-assessment were as follows: $X_{21} = 11$, $X_{22} = 14$, $X_{23} = 13$, and $X_{24} = 28$. Therefore, his change in performance was: $P_{t2}-P_{t1} = 23-18 = 5$. This meant that Gerald was engaging more frequently in occupations likely to lead to achievement of his life mission and to contribute to melioration of global issues of concern. Similarly, his aggregated self-rating scores on the MAIIIOT changed as follows: $X_{21}-X_{11}=11-8=3$, $X_{22}-X_{12}=14-9=5$, and $X_{23}-X_{13}=13-7=6$. This change in self-rating indicated that Gerald felt that he was performing more frequently and with increased adequacy occupations central to his mission in life, and he was more satisfied with his performance.

As mentioned earlier, although we do not know the exact impact of Gerald's increased performance of occupations directed towards improving pertinent global issues, we can suspect that based on the recursive effect which is a characteristic of complex systems such as our eco-system, such impact (outcome) may be disproportionately large compared to the changes that he made in his performance patterns. In other words, the relationship between change in performance and the effect in the world is not linear. Rather, it may be characterized by a non-linear function X^n where "n" stands for his field of influence by virtue of people and other agents in the ecosystem with which he has direct contiguity. Thus, Gerald's change in performance by 5 points translates into 5^n points in effect.

The analogy for this multiplier effect of performance is a scene from the movie "pay it forward" by Warner Bros (McLaglen, M., Treisman, J., Reuther, S., Abrams, P., et al., 2000). In the movie, Trevor McKinney (Haley Joel Osmert) comes up with an idea for an extra credit class assignment where a person who benefits from a good deed would be required to "pay it forward" by doing favors to three other people rather than "paying it back" to the person who did him/her the favor. In the movie, the idea that starts as a class assignment by an 11 year old boy quickly becomes a major movement that benefits many people.

Some may see the principle of "pay it forward" and other similar notions as utopian. However, throughout history, it is the idealism of utopia that in time becomes a socially transforming reality. This was the case in the renaissance and enlightenment periods and other revolutions in human history that led to improvement of the condition of humankind on earth, as well as in the idea of the mysterious hand proposed by Smith (1776) that seemingly regulates commerce in a free market economy. In other words, utopian ideas should not be dismissed out of hand before being tried. My contention in this book is that the principle of amplification of effects of actions is what gives individuals such as Gerald great power to significantly influence global issues through individual occupational performance, even when he may feel as if he is small and powerless. The reader is now invited to follow Gerald's example and explore how he/she can change his/her lifestyle for the good of all humanity, our planet, and all life on it.

Reflection Exercise #7

After reading chapter 7, complete the following exercises:

1. After reviewing the exploratory study discussed in the chapter, write down how you can:
   a. Create a vision based on your experiences and empathy for other people to guide your actions so that they are oriented towards a purpose that transcends personal self-interest.
   b. Use your own experiences as a basis for developing a sense of connectivity and empathy for other people.
   c. Translate your sense of morality into actions that lead to helping bring about circumstances that would allow all people in the world to live in freedom and dignity.
   d. Educate yourself about issues that cause loss of freedom and dignity for some people in the world and the historical and contemporary contexts that make those issues worse or better.

2. Following the guidelines provided in the individual, occupation-based framework for resolving global issues of concern to humankind discussed in this chapter:
   a. Create and write down your own personal mission statement that will act as a vision guiding your future occupational performance directed towards ameliorating global issues of concern.
   b. Using the instruments presented in the chapter, assess the extent to which your current occupational performance is consistent with that mission statement.
   c. Establish short-term and long-term goals whose achievement would lead to achievement of the mission.
   d. Identify occupations whose regular performance (or change in performance) would lead to achievement of the established goals and therefore your life mission (vision). If you are reluctant to change any of your major occupations even though the change is indicated, explore the cause of this reluctance and examine how you can derive the benefits afforded by the occupation in other ways.

References

Abrams, I. (1997). *Heroines of peace: The nine Nobel women* [Electronic version]. Retrieved April 27, 2007, from http://nobelprize.org/nobel_prizes/peace/articles/heroines/index.html.

Burrow, R. (2006). The defining biography of King?: A review essay. *Encounter, 67*(3), 297-315.

Cable News Network. (2000a). The priesthood years: Rebel with a cause [Electronic version]. *CNN*. Retrieved April 27, 2007, from http://www.cnn.com/SPECIALS/1999/pope/bio/priesthood/

Cable News Network. (2000b). The papal years: Charisma and restoration: Divine leader [Electronic version]. *CNN*. Retrieved April 27, 2007 from http://www.cnn.com/SPECIALS/1999/pope/bio/papal/

Capra, F. (1999). *The tao of physics: An exploration of the parallels between modern physics and Eastern mysticism*. Boston: Shambhala.

Capra, F. (1996). *A new scientific understanding of living systems: The web of life*. New York: Anchor Books/Doubleday.

Christensen, J. (2000a). The early years: An unhappy childhood: Boyhood faith [Electronic version]. *Cable News Network*. Retrieved from http://www.cnn.com/SPECIALS/1999/pope/bio/early/

Christensen, J. (2000b). John Paul II: Conscience of the world: Even the Catholic Church can survive a great man [Electronic version]. *Cable News Network*. Retrieved April 27, 2007, from http://www.cnn.com/SPECIALS/1999/pope/legacy/

Colaizzi, P. F. (1978). Psychological research as the phenomenologist views it. In R. Valle & M. King (Eds.), *Existential phenomenological alternative for psychology* (pp. 48-71). New York: Oxford University Press.

Cresswell, J. W. (1998). *Qualitative inquiry and research design: Choosing among five traditions*. Thousand Oaks, CA: Sage.

Depoy, E., & Gitlin, L. N. (2005). *Introduction to research: Understanding and applying multiple strategies* (3rd ed.). St. Louis, MO: Mosby.

Dewey, J. (1981). The influence of Darwinism on philosophy. In J. J. McDermott (Ed.), *The philosophy of John Dewey* (pp. 31-41). Chicago: The University of Chicago Press.

Dewey, J. (1960). The need for a recovery of philosophy. In W. G. Muelder, L. Sears, & A. V. Schlabach (Eds.), *The development of American philosophy* (pp. 404-417). Boston: Houghton Mifflin Company.

Dyer, W. W. (2004). *The power of intention: Learning to co-create your world your way.* Carlsbad, CA: Hay House.

Eakman, A. (2007). Occupation and social complexity. *Journal of Occupational Science, 14*(2), 82-91.

Giorgi, A. (1985). *Phenomenology and psychological research.* Pittsburg, PA: Duquesne University Press.

Ikiugu, M. N. (2007). *Psychosocial conceptual practice models in occupational therapy: Building adaptive capability.* St. Louis, MO: Elsevier/Mosby.

Ikiugu, M. N. (2004a). Instrumentalism in occupational therapy: An argument for a pragmatic conceptual model of practice. *International Journal of Psychosocial Rehabilitation, 8,* 109-117.

Ikiugu, M. N. (2004b). Instrumentalism in occupational therapy: A theoretical core for the pragmatic conceptual model of practice. *International Journal of Psychosocial Rehabilitation, 8,* 151-163.

Ikiugu, M. N. (2004c). Instrumentalism in occupational therapy: Guidelines for practice. *International Journal of Psychosocial Rehabilitation, 8,* 165-179.

Ikiugu, M. N., Anderson, L., & Anderson, W. (2007). *Occupational science in the service of GAIA: A study of the impact of human occupational behavior on global issues of our time.* Submitted for Publication.

Ikiugu, M. N., & Rosso, H. M. (2005). Understanding the occupational human being as a complex, dynamical, adaptive system. *Occupational Therapy in Health Care, 19*(4), 43-65.

Pope John Paul II. (2005a, January 29). *Address of pope John Paul II to members of the tribunal of the Roman Rota* [Electronic version]. Retrieved April 27, 2007, from http://www.vatican.va/holy_father/john_paul_ii/speeches/2005/january/documents/hf_jp-ii...

Pope John Paul II. (2005b, January 10). *Address of his Holiness Pope John Paul II to the diplomatic corps accredited to the Holy See for the traditional exchange of New Year greetings* [Electronic version]. Retrieved April 27, 2007, from http://www.vatican.va/holy_father/john_paul_ii/speeches/2005/january/documents/hf_jp-ii...

Pope John Paul II. (2004, January 12). *Address of his Holiness Pope John Paul II to the diplomatic corps accredited to the Holy See for the traditional exchange of New Year greetings* [Electronic version]. Retrieved April 4, 2007, from http://www.vatican.va/holy_father/john_paul_ii/speeches/2004/january/documents/hf_jp-ii...

Pope John Paul II. (1994). *Discourse to the plenary assembly of the Pontifical Council for Culture* [Electronic version]. Retrieved April 27, 2007, from http://www.vatican.va/holy_father/john_paul_ii/speeches/1996/documents/hf_jp-ii_spe_18...

Pope John Paul II. (1983). *Discourse to the plenary assembly of the Pontifical Council for Culture* [Electronic version]. Retrieved April 27, 2007, from http://www.vatican.va/holy_father/john_paul_ii/speeches/1996/documents/hf_jp-ii_spe_18...

van Kaam, A. (1959). A phenomenological analysis exemplified by the feeling of being really understood. *Individual Psychology, 15*, 66-72.

Kahakalau, K. (2004). Indigenous heuristic action research: Bridging Western and indigenous research methodologies. *Hulili: Multidisciplinary Research on Hawaiian Well-Being, 1*(1), 19-33.

Kennedy, G. (1996). John Dewey: Introduction. In M.H. Fisch (Ed.), *Classic American philosophers* (pp. 327-335). New York: Fordham University Press.

King, M. L. (1963a, August 28). *A call to conscience: The landmark speeches of Dr. Martin Luther King, Jr.: "I have a dream," Address delivered at the march on Washington for jobs and freedom* [Electronic version]. The Estate of Martin Luther King, Jr. Found at file://D/address_at_march_on_washington.htm(1-4of4)[7/17/20013:08:35PM.

King, M. L. (1963b, April 16). *Letter from Birmingham jail* [Electronic version]. The Estate of Martin Luther King, Jr. Found at www.kingpapers.org.

Lincoln, Y. S., & Guba, E. G. (1985). *Naturalistic inquiry*. Beverly Hills, CA: Sage.

Live Earth. (2007). *I pledge*: Retrieved July 7, 2007, from http://www.liveearth.org -SaveOurSelves-M...

van Manen, M. (1984). Practicing phenomenological writing. *Phenomenology and pedagogy, 2*(1), 36-69.

Maslow, A. H. (1970). Motivation and personality (2nd ed.). New York: Harper & Row.

Matta, D. S. (2004). *People of greatness: Their shared qualities and life experiences*. Master's Thesis, Trinity Western University, Ottawa, ON, Canada.

Mandela, N. (2005). *In full: Mandela's poverty speech* [Electronic version]. British Broadcasting Corporation News. Retrieved April 24, 2007, from http://news.bbc.co.uk/1/hi/uk_politics/4232603.stm.

Mandela, N. (1994, May 10th). *Inaugural address speech by Nelson Mandela* [Electronic version]. Retrieved April 24, 2007, from http://www.famousquotes.me.uk/speeches/Nelson_Mandela/index.htm.

Mandela, N. (1993). *Nelson Mandela: Nobel lecture* [Electronic version]. Retrieved April 24, 2007, from http://nobelprize.org/nobel_prizes/peace/laureates/1993/mandela-lecture.html.

Mandela, N. (1964). I am prepared to die [Electronic version]. *The History Place: Great Speeches Collection*. Retrieved April 24, 2007, from http://www.historyplace.com/speeches/mandela.htm.

Mark, J. (2002). *Nelson Mandela's speech* [Electronic version]. Retrieved April 24, 2007, from http://jmm.aaa.net.au/articles/4564.htm.

McLaglen, M., Treisman, J., Reuther, S., Abrams, P., Levy, R. L., Carson, P. et al. (Producers), & Leder, M. (Director). (2000). *Pay it forward* [Motion picture]. (Available from Warner Bros, Burbank, CA, USA).

Moustakas, C. (1990). *Heuristic research: Design, methodology and application*. Newbury Park, CA: Sage.

Nobel Foundation. (1979). *Mother Teresa: The Nobel peace prize 1979* [Electronic version]. Retrieved April 27, 2007, from http://nobelprize.org/nobel_prizes/peace/laureates/1979/teresa-bio.html.

Norwegian Nobel Committee. (1979). *The Nobel peace prize 1979* [Electronic version]. Retrieved April 27, 2007, from http://nobelprize.org/nobel_prizes/peace/laureates/1979/press.html.

O'Brien, D. (2007, February 14). *The seeds of wellness in the roots of our identity*. A Plenary Session presented at the 3rd Biennial Conference of the Behavioral Health and Safety for Rural America, Sioux Falls, South Dakota.

Patterson, G. J., & Zderad, L. T. (1976). *Humanistic nursing*. New York: Wiley.

Peirce, C. S. (1955). The fixation of belief. In J. Buchler (Ed.), *Philosophical writings of Peirce* (pp. 5-22). New York: Dover.

Reilly, M. (1962). Occupational therapy can be one of the great ideas of 20th century medicine. *American Journal of Occupational Therapy, 16*, 1-9.

Rogers, C. R. (1961). *On becoming a person*. Boston: Houghton Mifflin.

Rogers, C. R. (1950). A current formulation of client-centered therapy. *Social Service Review, 24*, 442-450.

do Rozario, L. (1997). Shifting paradigms: The transpersonal dimensions of ecology and occupation. *Journal of Occupational Science, 4*(3), 112-118.

Sadler, T. D. (2004). Moral sensitivity and its contribution to the resolution of socio-scientific issues. *Journal of Moral Education, 33*(3), 339-358.

Sames, K. M. (2005). *Documenting occupational therapy practice.* Upper Saddle River, NJ: Pearson/Prentice-Hall.

Speziale, H. J., & Carpenter, D. R. (2003). *Qualitative research in nursing: Advocating the humanistic imperative.* Philadelphia: Lippincott Williams & Wilkins.

Stenersen, O. (2004). *The humanitarian Nobel peace prizes* [Electronic version]. Retrieved April 27, 2007, from http://nobelprize.org/nobel_prizes/ peace/articles/stenersen/index.html.

Streubert, H. J. (1991). Phenomenological research as a therapeutic initiative in community health nursing. *Public Health Nursing, 8*(2), 119-123.

Mother Teresa. (1979, December 11). *Nobel lecture* [Electronic version]. Retrieved April 27, 2007, from http://nobelprize.org/nobel_prizes/peace/ laureates/1979/teresa-lecture.html.

Wilcock, A. (2006). *An occupational perspective of health* (2nd ed.). Thorofare, NJ: Slack.

Conclusion

In this book, it has been suggested that occupational science can contribute to the development of a fresh approach to healing our planetary ecology, including the earth's rocky crust, atmosphere, and biosphere [what Lovelock (1979, 2006) and Lovelock and Margulis (1974) refer to as the Gaia]. Gaia, as Lovelock and others argue, has been seriously injured by human pernicious activities that have resulted in destruction of the earth's vegetation cover (Gaia's skin), and pollution that has contaminated the constitution of atmospheric gasses (Gaia's respiratory system). It was pointed out that some of the symptoms of her ailment include decreased ability to sustain life, and warming [which Lovelock (2006) compares to human fever indicating illness] with subsequent increase in the frequency and intensity of natural calamities such as hurricanes, flooding, draughts, etc. The related problems that afflict humanity include diseases, poverty, corruption and government/institutional failures, war, material inequalities, and overpopulation.

It was proposed that the above listed problems are to a large extent a result of human activity as individuals pursue their daily occupations in the areas of self-maintenance, productivity, leisure pursuits, war, and sexual activity. Therefore, it was argued that the solution to these issues should be in part through a change in lifestyles of individuals as they pursue daily occupations. Furthermore, it was suggested that since occupational science is a basic scientific discipline whose specialization is illumination of the phenomenon of human occupational performance, it can provide useful insights that can be used to guide development of a fresh approach to facilitate lifestyle change among individuals as they pursue daily occupations so that their actions help improve global issues of concern and heal the earth.

Subsequently, a conceptual framework to facilitate change in individual occupational lifestyles, based on constructs derived from occupational science and occupational therapy was proposed. The constructs used in the framework were supported by findings from a small exploratory study that was conducted examining some of the characteristics of individuals who have impacted the world positively in a significant way. The framework (or conceptual model) consists of 4 steps: self-education so that one is informed about pertinent global issues of concern to humanity and to our planet and all life on it; mission establishment and self-assessment to determine the extent to which one's current occupational performance is consistent with melioration of global issues and healing of the injured earth; action; and consequence appraisal.

Suggestions for Implementation and Future Research

The framework presented in the book has not been empirically tested. It is therefore proposed that such testing be conducted through the following steps: First, a pilot study should be conducted to determine if the model could be actually useful in facilitating desired change in individual occupational lifestyles. The pilot study would be in form of a quasi-experiment, for example with college students, consisting of a pre-test, intervention, and post-test, with a control group. The pre-test/post-test data would be gathered using the instruments proposed in the model (the DOIs and MAIIIOT) [see chapter 7], and the intervention would consist of introducing students in the experimental group to the model's guidelines regarding how to educate themselves, formulate a mission statement and assess themselves, establish occupational lifestyle change goals, and evaluate performance consequences of instituted changes. Such a pilot study would determine change in occupational performance patterns, perceived frequency, adequacy, satisfaction with performance, and belief in ability to perform requisite occupations with desired frequency and adequacy.

Once it is determined through the pilot study that the model's guidelines are effective in facilitating occupational performance change in the desired direction (where occupational performance becomes more consistent with melioration of global issues of concern), the study would be replicated in a small community using the participatory action research approach. If the model is found to be useful in facilitating desired change in occupational performance among participants in a small community, it is recommended that a series of workshops be conducted. In these workshop participants would be exposed to the model's guidelines and then follow-up assessments would be conducted to determine if exposure results in occupational performance lifestyle changes.

References

Lovelock, J. E. (2006). *The revenge of the Gaia: Earth's climate crisis & the fate of humanity.* New York: Basic Books.

Lovelock, J. E. (1979). *Gaia: A new look at life on earth.* Oxford: Oxford University Press.

Lovelock, J. E., & Margulis, L. (1974). Atmospheric homeostasis by and for the biosphere: The gaia hypothesis. *Tellus, 26,* 2-9.

APPENDIX

Occupational Science in the Service of Gaia: A Study of the Impact of Human Occupational Behavior on Global Issues of Our Time

Questionnaire No. _____

The purpose of this study is to find out the extent to which individuals are aware of how their occupational choices shape world issues such as poverty, environmental destruction, climate change, etc. You have been chosen to participate because your expertise puts you in a position to provide useful information pertinent to this inquiry. Your participation is crucial to the success of the study. Please respond to the items in this questionnaire to the best of your ability. Thank you for participation!

PART ONE:
OCCUPATIONAL PERFORMANCE IN THE LAST TWO DAYS

1. The questions on this page ask you to list the activities in which you have participated during the past two days. You only need to offer a brief series of words or short sentences describing what you did. While it may be difficult to remember every activity that you performed every hour of the day, please do the best that you can. An example of a response may look like this: asleep, ate breakfast, read the newspaper, rode bicycle to and from work, worked 8 hours, had dinner with a friend, worked on the computer, went to bed. Think in terms of activities that you performed every hour from about 6:00 AM to midnight.

List all the activities in which you participated yesterday:

_____

_____

List all the activities in which you participated 2 days ago:

_____

_____

PART TWO:
ATTITUDE TOWARDS GLOBAL ISSUES
AND AWARENESS OF ONE'S IMPACT ON THEM

2. Now, we would like you to think about what may be considered to be some causes of poverty. The following list includes statements about what some people think are the primary causes of poverty. Please, read each statement and place a cross (X) on the appropriate box to indicate if you strongly agree, agree, somewhat agree, neither agree nor disagree, somewhat disagree, disagree, or strongly disagree with the following statement.

People are poor because:

	Strongly Disagree	Disagree	Somewhat Disagree	Neither Agree nor Disagree	Somewhat Agree	Agree	Strongly Agree
a. They are unable to manage their money							
b. They waste their money							
c. They do not work hard enough							
d. They are exploited by those with more money or resources.							
e. They live in an unjust system.							
f. Wealth is unfairly distributed in our society							
g. They lack opportunities to better themselves							

h. They have bad luck						
i. It is their fate						
j. They choose to be that way.						

3. Now, please read a series of statements about global problems. After reading each statement, please indicate if you strongly agree, agree, somewhat agree, neither agree nor disagree, somewhat disagree, disagree, or strongly disagree that human beings are responsible for each of these issues.

In my opinion, people are to a large extent responsible for:

	Strongly Disagree	Disagree	Somewhat Disagree	Neither Agree nor Disagree	Somewhat Agree	Agree	Strongly Agree
Unequal distribution of wealth, such as money or resources.							
Global Warming							
Diseases							
Corruption and failure of government							
Overpopulation							

PART THREE:
YOUR PERCEPTION OF THE EXTENT TO WHICH YOUR PERFORMANCE AFFECTS GLOBAL ISSUES

3. In the table below, list at least three of the activities that you identified in the Daily Occupational Inventory in Item#1. In your perception, your participation in each of those activities had what effect on the following global issues? Use the scale provided below to rate your perception.

0   Made the issue much worse
1   Made the issue a little worse
2   Did not affect the issue
3   Made the issue a little better
4   Made the issue much better

LIST YOUR DAILY ACTIVITIES (OCCUPATIONS) HERE (as entered in item# 1 above)	Poverty	Occurrence and consequences of diseases (e.g. HIV/AIDS, heart diseases, etc.	Material inequalities	Corruption and failure of government and/or social institutions	Over-population	Environmental destruction
Example: Driving to work	2	1	2	2	2	1

4. If you could, would you change the way you perform any or all of the activities you listed in item# 1 so as to contribute towards improvement of any or all of the global issues enumerated in item #3?
(Check one) __ Yes __ No __ Undecided

5. If you chose "yes" in answer to the above question, please indicate which of the activities you listed that you would change and how. If there are activities in which you did not engage in the last 2 days but which you think would help you impact the listed global problems, list them as well and explain how they would empower you to be more effective in positively affecting the issues in question.

Example:

My listed activity was: My proposed change is:
Driving to work   To use public transportation instead of driving so as to reduce environmental destruction

_____

_____

_____

PART FOUR:
DEMOGRAPHICS

6. In what year were you born? _____

7. Relationship Status
PLEASE SELECT A RESPONSE BY PLACING A CROSS [X] ON THE
APPROPRIATE LINE
    a. Married _____
    b. Cohabiting _____
    c. In a Relationship ___
    d. Widowed _____
    e. Divorced _____
    f. Separated _____
    g. Single_____
    h. Other _____ (Explain)

8. Level of Education
PLEASE SELECT A RESPONSE BY PLACING A CROSS [X] ON THE
APPROPRIATE LINE
    a. Some College (Associate Degree or College Certificate) _____
    b. Graduated (Have a College Degree/Diploma) _____
    c. Some Graduate Education (no higher degrees) _____
    d. Graduate Education (Have a Higher Diploma, Master's degree, or
Doctoral Degree) _____

Note: The Daily Occupational Inventories (item number 3) were adopted
from Ikiugu (2007). The Inventory was first designed for use in research by
Ikiugu and Rosso (2005). The attributions of poverty in item # 4 were derived
from Shek (2004, pp. 278-279).

Endnotes

[i] I would like to apologize to Christins for this critique of what I perceive to be the failures of contemporary Christianity. I do not mean to suggest that their religion is decadent or that it is plagued by unique problems that are not in other religions. I use Christianity as an example because it is the faith in which I was raised. Therefore, I am more familiar with its teachings and with issues that commonly face its adherents than I am with any other religion. Also, I want to point out that criticism of the shortcomings of Christianity as a religious institution is not meant in any way to be a criticism of individual Christians or a suggestion that religion is not useful in human lives. I know Christians who are very decent, caring individuals. I also know some individuals for whom Christianity is a very positive influence in their lives. For some of them, their lives could have been recked were it not for their religious faith. Furthermore, spirituality is recognized in occupational therapy and occupational science as a very important human occupation (Law, Baptiste, Carswell et al., 1998; Law, Polatajko, Baptiste, & Townsend, 2002). Of course, as understood in occupational therapy and occupational science, spirituality as the essence of a human being is clearly distinguished from religion as a practice. However, religion is the means by which many people express their spirituality, and therefore, it is and should be an important occupational pursuit in the perspective of occupational therapy and occupational science.

[ii] At the time of this writing, at least 41 people had been reported dead in the Southern and Mid-Western regions of the USA as a result of persistent heat waves in the summer of 2007 (Rucker, 2007)

[iii] Again, in the summer of 2007, many Mid-Southern states of the USA were inundated by flood waters and thousands of people in those states lost their homes. Many had to be rescued by boats and helicopters from flood waters that were as high as over 8 feet in many places. CNN reported that this was the worst flooding in the USA history since reliable record keeping began.

[iv] One may argue that outsourcing is not necessarily bad. In fact, it may be good globally because it provides jobs for people in poor countries who may otherwise have no jobs. It can therefore be seen more or less as a global wealth re-distribution measure. The problem with this argument is that the practice of outsourcing encourages exploitation of desperately destitute people living in the poorest regions of the world. Workers in those circumstances are paid only a pittance for their labor, making it difficulty for them to rise above conditions of abject poverty. The practice also facilitates establishment of inhuman working conditions, such as sweatshops where poor employees work long hours, sometimes without even bathroom breaks.

For example, in 2002 I visited El Salvador with a mission group sponsored by the Jesuits consisting of University of Scranton employees and faculty members. Many El Salvadorans told us stories about the horrors of working in poorly ventilated sweatshops belonging to some well known leading textile corporations. They told us that sometimes, they worked for periods of up to 16 hours at a time, for very little pay, with very few breaks in between. Therefore, in my view, outsourcing would humanely work only if some kind of international standards are established to mandate how employees should be treated regardless of where they are in the world. We have standards prohibiting corporate use of child labor. Why can't similar standards based on the need to pay employees "living wages" be agreed upon by the international community? It may be possible to calculate what percentage of any country's GDP a person needs to earn in wages in order to live a relatively decent life. International standards can then be established mandating that irrespective of where a corporation does business, it has to pay its employees wages and benefits at least equivalent to the percentage of that country's GDP determined to constitute "living wages", in addition to ensuring that other standards guiding how to treat employees are followed. Such an approach would decrease the incentive to outsource jobs due to the ability to exploit workers in poor regions of the world. It would also increase productivity because workers would be aware that minimum wages are dependent on the overall national wealth. Higher GDP would mean higher minimum wages and vice-versa. Workers would therefore have a stake in increasing national productivity, which would be consistent with Smith's doctrine of "self-interest" as a basis for national wealth generation (this principle should be highly attractive to adherents of unmitigated capitalism).

[v] As the second draft of the manuscript for this book was being prepared, Al Gore and colleagues (Ollison, 2007) organized a worldwide event that they referred to as the "Live Earth Concert" (which took place in Australia, China, South Africa, Brazil, Germany, and USA). It was expected that about 2 billion people would participate in the event. Apart from raising awareness about global warming and climate change, it was hoped that participants would sign a pledge in which they would promise to alter their lifestyles in order to impact the planet positively. Some of the actions in the pledge included advocating for governments to join an international treaty to curb activities that contributed to climate change, reducing personal contribution to emission of carbon dioxide and other greenhouse gasses, pressuring governments to cease constructing new energy producing plants that contributed to accumulation of greenhouse gasses in the atmosphere, committing to increasing personal energy efficiency at home, workplace, school etc., planting new trees, and buying goods from businesses that demonstrated commitment to resolution of the problem of global warming and climate change (Live Earth, 2007). These proposals placed the responsibility of effecting global change on individuals. This approach was similar to what is proposed in this book as a way of helping resolve many of the

global issues related to global warming and climate change such as poverty, material inequalities, overpopulation, corruption and government/institutional failure, etc. (see chapter 7for details).

Printed in the United Kingdom
by Lightning Source UK Ltd.
132579UK00001B/59/P